Contents

1 It's geography

geog.1

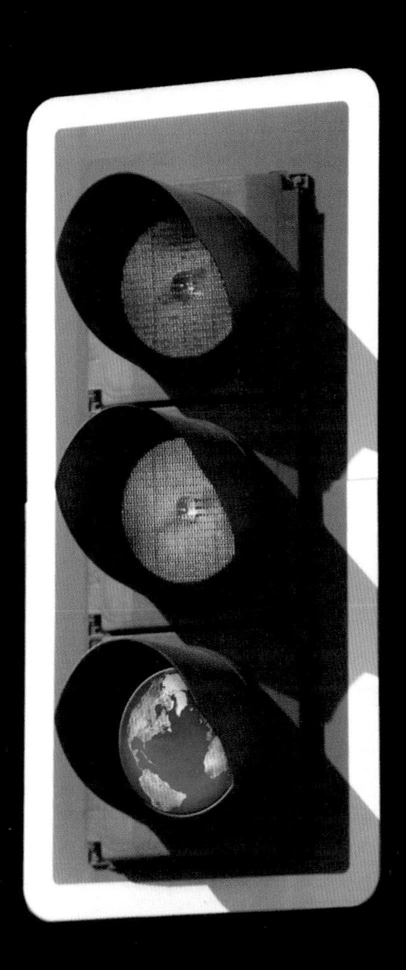

geography for key stage 3

<rosemarie gallagher><richard parish><janet williamson>

OXFORD

Great Clarendon Street, Oxford OX2 6DP

Oxford University Press is a department of the University of Oxford. It furthers the University's objective of excellence in research, scholarship, and education by publishing worldwide in

Oxford New York

Auckland Cape Town Dar es Salaam Hong Kong Karachi Kuala Lumpur Madrid Melbourne Mexico City Nairobi New Delhi Shanghai Taipei Toronto

With offices in

Argentina Austria Brazil Chile Czech Republic France Greece Guatemala Hungary Italy Japan Poland Portugal Singapore South Korea Switzerland Thailand Turkey Ukraine Vietnam

First published 2000
Second Edition 2005
Third Edition 2008

British Library Cataloguing in Publication Data

Data available

ISBN 978-0-19-913493-9

10 9

Printed in Singapore by KHL Printing Co. Pte. Ltd

Paper used in the production of this book is a natural, recyclable product made from wood grown in sustainable forests. The manufacturing process conforms to the environmental regulations of the country of origin.

Acknowledgements

The publisher and authors would like to thank the following for permission to use photographs and other copyright material:

4 PLI/Science Photo Library; 6 GSFC/NASA; 7l Steve Bloom Images/Alamy, 7c WoodyStock/Alamy, 7r Leland Bobbè/Corbis UK Ltd.; 8tl Douglas Peebles Photography/Alamy, 8bl Stock Connection Blue/Alamy, 8tc MONSERRATE SCHWARTZ/Alamy, 8bc Chad Ehlers/Alamy, 8tr RoseMarie Gallagher, 8cr PLI/Science Photo Library, 8br Bjorn Svensson/Alamy; 9l M Stock/Alamy, 9r Steve Bloom Images/Alamy; 10tl Nik Wheeler/Corbis UK Ltd., 10bl Bryan & Cherry Alexander Photography/Alamy, 10tr Iconotec/Alamy, 10br 2007 Mike Goldwater/Alamy, 10 PLI/Science Photo Library; 11l Raine Vara/Alamy, 11r Neil Rabinowitz/Corbis UK Ltd.; 12tl PLI/Science Photo Library, 12bl Skyscan Photolibrary/Alamy, 12tr David Chapman/Alamy, 12br Louie Psihoyos/Corbis UK Ltd.; 13l Popperfoto, 13r Popperfoto; 14tl PLI/Science Photo Library, 14cl PLI/Science Photo Library, 14bl John Stephenson/LGPL/Alamy, 14tr JRC.Inc./Alamy, 14cr W Cross/Skyscan, 14br Martin Sookias/Oxford University Press; 16 Oxford University Press; 17 Viviane Moos//Corbis UK Ltd.; 18 Oxford University Press; 19 educationphotos.co.uk; 20 Oxford University Press; 22 Getmapping PLC; 24 NRSC//Skyscan; 25 educationphotos.co.uk; 27t educationphotos.co.uk, 27b Marktime Travel & Leisure Ltd.; 32 Oxford University Press; 34 Skyscan; 37bl Ric Ergenbright//Corbis UK Ltd., 37tc C. Moore//Corbis UK Ltd., 37bc Design Pics Inc./Alamy, 37br Charles & Josette Lenars//Corbis UK Ltd., 37t Christophe Loviny//Corbis UK Ltd.; 38 Martyn F. Chillmaid; 39 Simmons Aerofilms; 42bl RoseMarie Gallagher, 42cr Bucks Herald, 42br RoseMarie Gallagher, 42t RoseMarie Gallagher, 42bcr RoseMarie Gallagher, 42blc RoseMarie Gallagher; 43l RoseMarie Gallagher, 43r RoseMarie Gallagher; 46 London Aerial Photo Library; 47l Bucks Herald, 47c Aylesbury Vale District Council; 48l RoseMarie Gallagher, 48r RoseMarie Gallagher; 51 Careys New Homes; 52t Mary Evans Picture Library, 52b Oxford University Press; 56bl Bluewater, Kent, 56br Tony Kyriacou//Rex Features, 56t Andy Drysdale//Rex Features; 57 Bluewater, Kent; 58 Michael Keller/Corbis UK Ltd.; 59l vario images GmbH & Co.KG/Alamy, 59r Ferruccio/Alamy; 60tl Oxford University Press, 60cl Images of Birmingham/Alamy, 60bl John Miller/Robert Harding/Corbis UK Ltd., 60bc Mike Hewitt/Getty Images, 60tr Nigel Stollery/Alamy, 60cr Oxford University Press, 60br Adrian Sherratt/Alamy; 62 Oxford University Press; 63l J. Allan Cash Photo Library, 63r Barnaby//Mary Evans Picture Library; 65 Oxford University Press; 69cl Barnaby's//Mary Evans Picture Library, 69bl Barnaby's Picture Library, 69cr Barnaby's//Mary Evans Picture Library, 69br Getty Images, 69t Barnaby's Picture Library; 71l Paul Tomkins/VisitScotland/Scottish Viewpoint, 71r London Aerial Photo Library//Corbis UK Ltd.; 72tl C.CAVU/Everett/Rex Features, 72bl Alexander Caminada/Rex Features, 72tr Jim Wileman/Alamy, 72br Andreas Lander/Dpa/Corbis UK Ltd.; 74tl Bettmann//Corbis UK Ltd., 74cl Tony O'Keefe//Oxford University Press, 74bl ian woolcock/Alamy, 74tr Corel//Oxford University Press, 74cr Tony O'Keefe//Oxford University Press, 74br Eddie Ryle-Hodges/Edifice//Corbis UK Ltd., 74bcl Oxford University Press, 74bcr Adam Woolfit//Corbis UK Ltd.; 75l Chris Bland/Eye Ubiquitous//Corbis UK Ltd., 75r Stan Gamester/Photofusion Picture Library//Alamy; 76l vario images GmbH & Co.KG/Alamy, 76r Giles Moberly/Alamy; 78 Ron Niebrugge/Alamy; 80 Pixland/Corbis UK Ltd.; 83 AirFotos; 84 Art Directors & Trip Photo Library; 85 Landform Slides; 86t ICCE Photolibrary, 86b David Norton/Alamy; 87 Heather Angel//Natural Visions; 89tl Mark Boulton/Alamy, 89tr South West News Service, 89b Paul Glendell/Alamy; 90 David Goddard/Getty Images; 92 Christopher Furlong/Getty Images; 93 Peter Finnigan/Tewkesbury Historical Society; 97 Matt Cardy/Getty Images; 98tl Wildscape/Alamy, 98bl (c) Crown copyright 2007/John Ffoulkes/Met Office 98tc Kirsty Wigglesworth/Associated Press/PA Photos, 98bc Daniel Berehulak/Getty Images, 98tr Matt Cardy/Getty Images, 98br GPA Images/Rex Features, 98/99 Mark Boulton/Alamy, 99tl Peter Nicholls/AFP/Getty Images, 99tc Matt Cardy/Getty Images, 99bc Mark Boulton/Alamy, 99tr Christopher Furlong/Getty Images, 99br Paul Glendell/Alamy, 99br Matt Cardy/Getty Images; 101 Daniel Berehulak/Getty Images; 103t Mark Boulton/Alamy; 106tl Barbara Walton/Epa/Corbis UK Ltd., 106bl Richard Wainscoat/Alamy, 106tr Gary L. Tong/Newsport/Corbis UK Ltd., 106cr Henrik Trygg/Corbis UK Ltd., 106br Tim De Waele/Corbis UK Ltd.; 107tl Olivier Maire/Keystone/Corbis UK Ltd., 107bl Rahat Dar/Epa/Corbis UK Ltd., 107tr Karl Weatherly/Corbis UK Ltd., 107br How Hwee Young/Epa/Corbis UK Ltd.; 108tl Oxford University Press, 108cl RoseMarie Gallagher, 108bl Laurence Griffiths/Getty Images, 108tc Liverpool Football Club, 108tr Liverpool Football Club, 108c Liverpool Football Club; 109 Liverpool Football Club; 111t London Aerial Photo Library, 111b Liverpool Football Club; 112l Ben Radford/Corbis UK Ltd., 112c Ben Radford/Corbis UK Ltd., 112r Ed Kashi//Network Photographers Ltd; 113 Popperfoto; 114tl Jason Hawkes/Corbis UK Ltd., 114bl London 2012, 114tr Philip Wolmuth/Alamy, 114br Premium Stock Photography GmbH/Alamy; 115 London 2012; 116t Nick Duff Davies/Alamy, 116b Daniel Berehulak/Getty Images; 117t Kevin Reece /Icon Smi/Corbis UK Ltd., 117b Ker Robertson/Getty Images; 118 Bernard Edmaier/Science Photo Library; 120l Corel//Oxford University Press, 120r KTB - Archive of the GeoForschungZentrum Potsdam; 121l Oxford University Press, 121r Oxford University Press; 122 Paul A. Souders/Corbis UK Ltd.; 125 (Source: World Ocean Floor map by Bruce C. Heezen and Marie Tharp, 1977. Copyright © 1977 Marie Tharp. Reproduced by permission of Marie Tharp, 1 Washington Ave., South Nyack, NY10960)//Marie Tharp; 127 AP Photo; 128 Zahid Hussein/Reuters/Corbis UK Ltd.; 129t Aamir Qureshi / Pool/Epa/Corbis UK Ltd.; 129b Farooq Naeem/AFP/Getty Images; 130 JOHN RUSSELL/AFP/Getty Images; 131tl DigitalGlobe/Getty Images, 131tr Digital Globe/Zuma/Corbis UK Ltd., 131b ullstein - Mehrl/Still Pictures; 132l J. Allan Cash Photo Library, 132r Lyn Topinka/United States Department of the Interior//U.S. Geological Survey; 133t Rex Features, 133b Katz Pictures; 134l Marco Fulle/Osservatorio Astronomico/Trieste, 134r NASA; 135 Kevin West/Liaison/Getty Images; 136tl Yannis Kontos/Sygma//Corbis UK Ltd., 136cl Patrick Robert/Sygma//Corbis UK Ltd. 136bl Alan Andrews Photography//Alamy, 136tc Peter Turnley//Corbis UK Ltd., 136cc Jacques Langevin/Sygma//Corbis UK Ltd., 136bc Thomas J. Casadevall/United States Department of the Interior//U.S. Geological Survey, 136tr Peter Frischmuth//Still Pictures, 136cr Dimas Ardian/Getty Images, 136br Kimimasa Mayama/Reuters/Corbis UK Ltd.; 137 NASA

The Ordnance Survey map extracts on pp. 31, 32, 45, 97 and 111 are reproduced with the permission of the Controller of Her Majesty's Stationery Office © Crown Copyright.

Illustrations are by James Alexander, Martin Aston, Barking Dog Art, Jeff Bowles, Matt Buckley, Stefan Chabluk, Michael Eaton, Jeff Edwards, Hardlines, Jill Hunt, Tim Jay, David Mostyn, Tim Oliver, Olive Pearson, Colin Salmon, Martin Sanders, Jamie Sneddon, and Tony Wilkins.

The publisher and authors would like to thank the many people and organizations who have helped them with their research. In particular: Clare Locking at the Planning Department, Aylesbury Vale District Council; Caroline Jiggins at Aylesbury Grammar School; Janice Halsey at Aylesbury Tourist Information Centre; The Environment Agency; and Liverpool Football Club. Special thanks to Alex 'Walter' Middleton.

We would like to thank our excellent reviewers, who have provided thoughtful and constructive criticism at various stages in the development of this course: Phyl Gallagher, Caroline Jiggins, John Edwards, Anna King, Katherine James, Paul Apicella, David Weeks, Richard Farmer, Roger Fetherston, Philip Amor, Louise Ellis, Kathy Fairchild, Janet Wood, Andy Lancaster, David Jones, Bob Drew, Paul Bennett, John Hughes, and Michael Gallagher.

Every effort has been made to contact copyright holders of material reproduced in this book. Any omissions will be rectified in subsequent printings if notice is given to the publisher.

The big picture

Welcome to *geog.1*, the first book of the *geog.123* course.

This course is all about planet Earth, and how it is changing. These are the big ideas behind the course:

- Our planet is always changing.
- Natural processes are changing it. For example the action of rivers, and hot currents inside the Earth.
- Humans are also changing it. We have spread over most of the land. We farm it, build on it, and dig it up for metals and fuel.
- We have carved the Earth up into nearly 200 countries. They are all different – but all depend on each other.
- We have made many mistakes. We've spoiled places, and wiped out species. Now experts say we are making the Earth warm up.
- We need to look after our planet properly.

Did you know?

- *The Earth has been here about 23 000 times longer than humans.*

Your goals for this course

By the end of this course, we hope you will be a good geographer! And that means you will:

- be interested in the world around you.
- understand that many processes, both natural and human, are changing and shaping the surface of the Earth.
- know what kinds of questions to ask, to find out about countries and places and people, and how things are changing.
- be able to carry out enquiries, to find answers to your questions.
- have the other key skills (such as map reading) that a geographer needs. Your teacher will tell you which ones.
- think geography is just brilliant !

What if ...

- *...we were taken over by creatures from another planet?*

Your chapter starter

Page 4 shows a planet. Which one?

Where in space is it?

What's keeping it there?

Who's on it?

What are they doing?

1.1 Hey, you over there!

In this unit, we do a quick survey of the planet you live on.

Planet Earth, your home

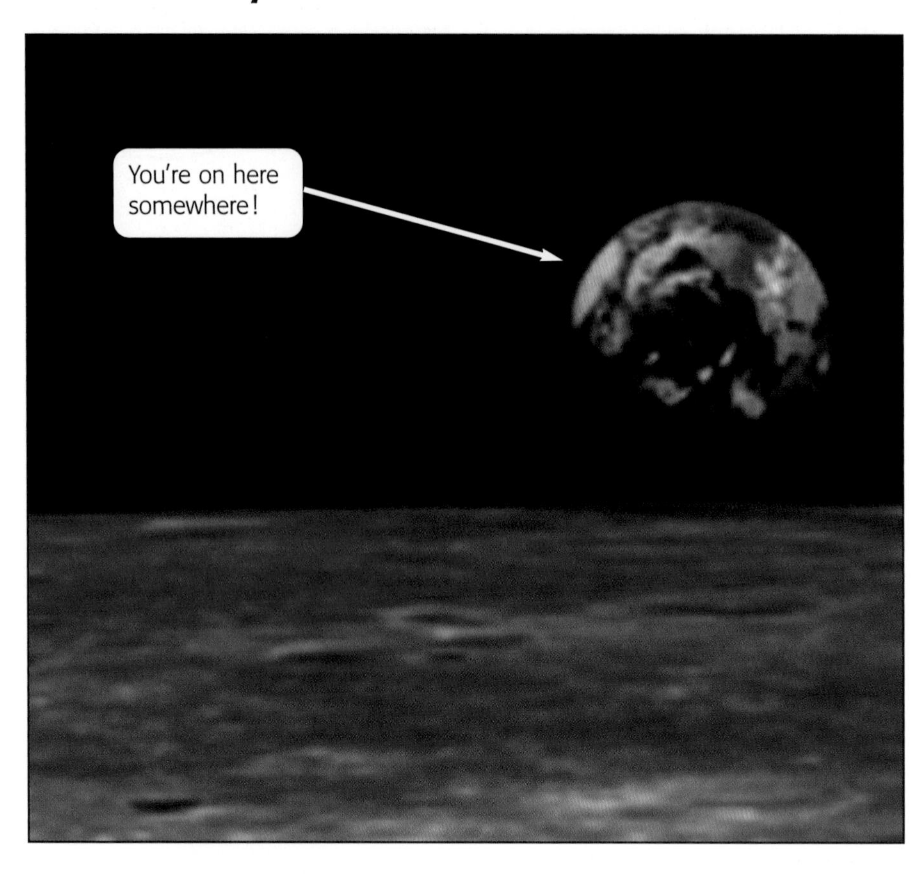

Planet Earth, seen from the moon.

You live on planet Earth.

- You are young. The Earth is very old. About 4.6 billion years old.
- It is held in space by a force between it and the sun. This force attracts objects to each other. It is called **gravity**.
- The Earth travels non-stop around the sun, taking you along too. (So while you sit here, reading this, you are on a cosmic journey.) It shoots along at 108 000 km an hour. One full orbit around the sun takes it a year.
- It spins as it goes. Here at the UK, it is spinning at about 1400 km an hour. One full spin takes a day.
- But you don't fly off as it spins – because you are held to it by gravity.

What's it like?

Your planet home is big – about 40 000 km around its middle.

It is made of rock, and, deep inside, two metals, iron and nickel.

Over two-thirds of it is covered by oceans. These are very deep in places. But compared with the size of the Earth, they are just a thin film of water.

Around the Earth is a layer of gas, that travels with it. This layer is called the **atmosphere**. It is over 100 km deep – but most of the gas lies in the lowest 30 km. We call it **air**. You breathe it in.

Full of life ...

From space, your planet looks cold and lonely. But it is full of life. There are around 1.8 million *known* species of living things on land and in the sea. Here are just three of them:

Elephants are our largest land animal. They have been around for millions of years. But now there are only about 600 000 left. (Why?)

We think the first jellyfish appeared more than 500 million years ago. Now there are over 200 kinds, and you find them in every ocean.

And this species has been around for only 200 000 years. There are over 6.7 billion of us – and the number is growing fast.

New species are being found all the time, on land and in the ocean. Some scientists say there may be up to 30 million species on the Earth. Some say 100 million.

... and always changing

From space, the Earth looks quiet and unchanging. Don't be fooled! It is changing all the time, in all kinds of ways. Because of:

- natural processes, and
- the actions of us humans

You can find out more about these changes in the next unit. And you'll look at them more closely later in your course!

Your turn

1 a What age are you?
  b So how many times have you been around the sun?
  c Suppose the school bus goes at 50 km an hour. How many times faster is your journey around the sun?

2 The Earth is 4.6 billion years old. 4.6 billion is ...?
  a 460 000 b 46 000 000 c 4 600 000 000

3 Now look at this diagram. It shows the Earth spinning.
  a It's dark at A. Why?
  b At which place, **A**, **B** or **C**, do you think the Earth's spin is slowest? See if you can explain why.
  c Can you feel the Earth spinning? See if you can give a reason.

4 Of all the facts about the Earth on these two pages, which one do you like best? Why?

5 Now write a space-mail to your friend DK3 on planet Dkvorak, saying what *you* think is great about planet Earth. You could invite her to visit.

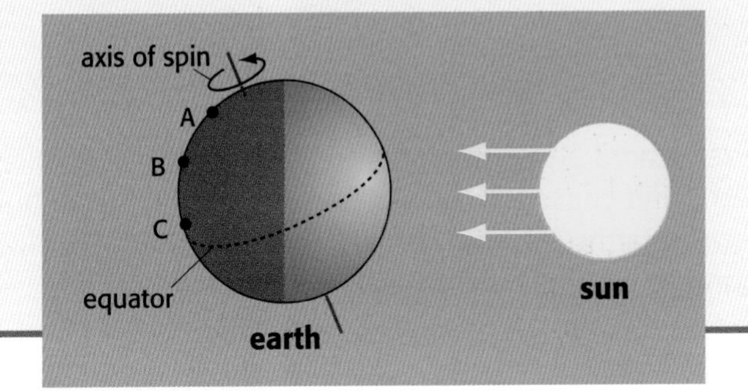

Our planet: always changing

Here you will find out how natural forces, and we humans, are changing our planet.

All change!

You might not notice that your planet is changing. But it is, all the time. It's being changed by natural forces. And by us.

1 Changed by natural forces

Deep inside the Earth, currents of hot soft rock are flowing – causing earthquakes and volcanoes, and even making mountains grow.

At the Earth's surface, other things are flowing: rivers, waves, glaciers, wind. They scrape and shape the land as they flow.

And everywhere, rock is being broken down to soil, in a process called **weathering**. (It is mainly due to the weather.)

2 Changed by us

The Earth is around 4.6 billion years old. That's 4 600 000 000 years old. Humans like us appeared only about 200 000 years ago. But we have made huge changes already.

We've cleared away most of the forests, and chased away wildlife, to set up farms. We've dug up the ground to get fuels and metals.

We have built villages, towns and cities. We have built roads all over. Now you'll find humans nearly everywhere.

We have divided the Earth into over 200 countries, and put borders between them. You may need a visa (a special pass) to get through.

▲ *We cause it – and it harms us. The wind carries air pollution all around the world.*

▲ *A mountain gorilla. Fewer than 650 are left on the Earth, thanks to hunting and the chopping down of forests.*

So are all these changes a problem?

Natural changes can cause us big problems. For example if an earthquake strikes our place, or a river floods it.

Human changes are causing even bigger problems. Like these:

- We have already killed off many species of plants and animals, by taking over their land, and by hunting them.
- Experts say we are making the Earth warmer, by burning so much **fossil fuel** – coal, oil and gas. This **global warming** will bring many disasters. Such as terrible storms, floods, famine.
- Many of the changes we make cause conflict, and even wars.
- We have created an unfair world. Many humans have plenty of everything. But many have almost nothing, not even enough to eat.

You'll find out more about these problems, and how we can solve them, in the rest of your course.

What if ...
... we killed off all the gorillas, and pandas, and tigers, and whales, and ...?

Your turn

1 Think about the natural forces that are changing the Earth.
   a Which of them do you think we can control?
   b Do you think they were at work *before* humans arrived on the Earth?
   c Choose one that you think is helping us, and say why.
   d These natural forces can spell danger for us. See how many of the dangers you can list.

2 Now think about where you live. Do you think any natural processes are changing your area? Do you notice any changes?

3 When humans appeared, the Earth was a wild place with thick forests. In what ways have we changed it? Write a list. See how much you can add to it.

4 Think about where you live. Is it being changed by humans? What changes are going on right now?

5 Look at the problems given in the last section above. Which do you think is the most serious? List them in order, as very short bullet points, the most serious first.

6 Now think about the questions on the right. What are your answers?

Why bother looking after it?

Whose planet is it anyway?

1.3 Your place on the planet

We all have a place on the planet. What makes our places different? That's what you will explore in this unit.

Everyone has a place

Everyone has a place on the planet – you, me, the King of Tonga. So let's have a look at some of our places.

This is Hassan's place. He is a Marsh Arab. He and his family live on an island of reeds in the marshes in southern Iraq. They go everywhere by boat.

This is Rana's place – a small village in Mali. Her dad is a farmer. She helps to look after the little ones. And walks everywhere. No other way to go!

This is Alona's winter place. Her family are reindeer herders, in Siberia in Russia. They move around the tundra with their reindeer, to the best grazing places.

Vitor's place – on the street in Recife, in Brazil. That's him front left. His mum died and he has no home to go to. He's made friends with the other street children.

Emi's place. She is Japanese, and lives in an apartment on the 31st floor, in Tokyo – the world's largest city. At night she looks down on the bright lights of Tokyo.

Sela's place. She lives in Tonga, a country of 169 islands, in the Pacific Ocean. It has 117 000 people. And like the UK, it has a Royal Family.

So what's *your* place like?

Your place is a tiny dot on the planet. Billions of people may never even have heard of it.

But to you it's special. You have memories of it, and images of it, and feelings about it. And it's home – at least for now.

Your turn

1 Look at the six photos of places.
 a Which place would you most like to spend time in? Why?
 b Which would you least like to spend time in? Why?
 c Which one looks least changed by humans?

2 Now choose one of the six places. (It need not be your favourite.) Imagine you are there, looking around you. Take your time. Relax. Now:
 a What can you see?
 b What can you hear?
 c What can you smell?
 d How do you feel about this place?

3 In geography, we always like to know *where* places are, on the planet.
 So turn to the map of the planet on pages 140 – 141.
 It shows the countries we've divided the planet into.
 Find the country where each photo was taken.
 Then work out which continent it is in.
 Write your answer like this:
 Place A is in Iraq in Asia.

4 Now, what about *your* place? Imagine you are standing outside where you live.
 a Which country are you in?
 b Which continent are you in?
 c What can you see around you?
 d What can you hear?
 e What can you smell? Anything?
 f How do you feel about this place?

5 When we are away from our place, we still 'see' it in our minds. We can 'see' our friends, family, pets, things we own, things we like to do.
 Close your eyes and think about your place for a few minutes.
 a What kinds of images came to your mind?
 b Describe one of them.

6 Can you name a place you think is better than your place? If yes, in what way is it better?

7 Do you think someone else from your place would give the same answers as you, to questions 4 – 6? Explain.

It's all geography!

This unit shows the kinds of topics you'll study in geography – and how being nosy will help!

Glorious geography

Geography is about everything that's going on, in places all over the planet, right at this moment.

The world can be confusing. But geography will help you make sense of it.

Dividing up geography

Geography is a big subject. So it helps to divide it up. For example into the three areas shown here.

Look at the kinds of things you will learn about, in each area.

Wow!

1 Physical geography – about what our planet is like

You'll learn about rivers, the sea, the coast, the weather, and climate. And about the natural dangers we face – such as earthquakes, volcanic eruptions, and floods.

2 Human geography – about how and where we live

You'll learn about the places we live in, and how they are changing; things we get up to – like work, sport, crime; other countries, and why many are poor; how we depend on each other; how we can make the world a fairer place.

3 Environmental geography – about how we affect our surroundings

You'll learn about how we pollute the land, air and water; how we kill off species; how we waste things; how we are warming the Earth up; and how we are learning to take more care.

So, get ready to geog!

The first step to being good at geography is: get nosy!

Use your eyes. Look around you. Look for clues. Ask questions that start with *Who, What, Where, How, Why, When* …

And have fun.

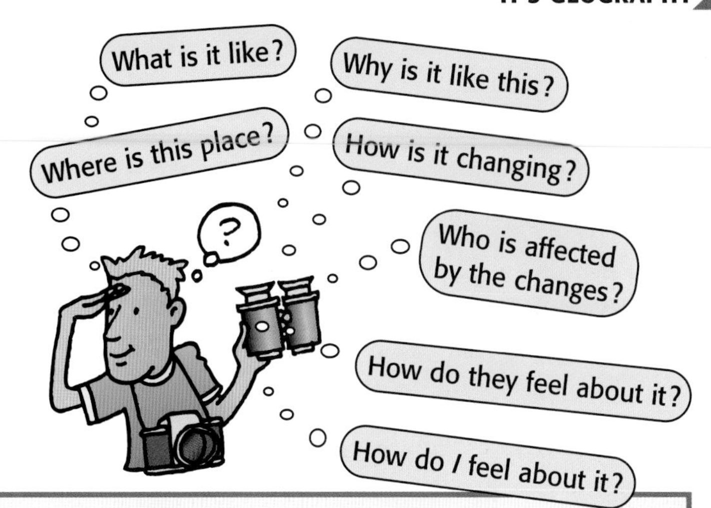

Your turn

1 Copy and complete:
Physical geography is about …
Human geography is about …
Environmental geography is about …

2 Which kind of geography is this topic?
- a how clouds form
- b which countries are crowded
- c protecting pandas
- d where trainers are made
- e caves
- f dumping rubbish in rivers

3 Photo **A** below shows people on holiday.
- a Why do you think they chose this place? List as many reasons as you can.
- b After each reason, write *(P)* if it's about physical geography, *(H)* if human, or *(E)* if environmental.

4 Time to get nosy! Study photo **B** for clues. Then answer these questions:
- a What is going on in the photo?
- b How did the place get to be like this?
- c Who do you think is responsible?

5
- a Now make up three new questions about photo **B**, and what's going on there. No silly ones! (Hint: *Who? What? Where? How? Why? When?*)
- b Ask your partner to try to answer them.

6 Compare the two photos.
- a Can you see any similarities?
- b Do you think there is any connection at all between the two scenes?

A

B

Making and mapping connections

Where is Walter?

On planet Earth, with over 6.6 billion other humans (that's 6 600 000 000), including you …

… in Europe, with over 809 million other humans (that's 809 000 000) …

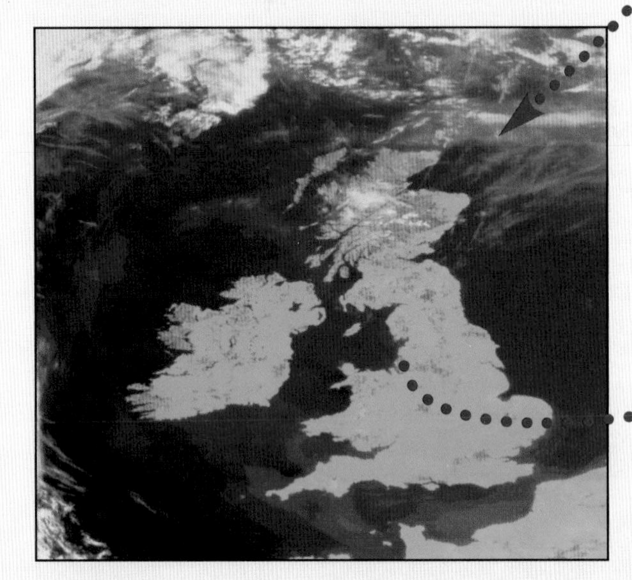

… in the British Isles, with about 65 million other humans …

… in Liverpool, with over 440 000 other humans …

… in number 181 Anfield Road, with 4 other humans …

… in this room, all alone.

The big picture

This chapter is all about maps, and how to use them. These are the big ideas behind the chapter:

- We humans are spread out all over the Earth – but we are connected to each other in many different ways.
- We use maps to show where we live on the Earth, and what places are like.
- There are many different kinds of maps.
- Using maps is a key skill for a good geographer. (That's you!)

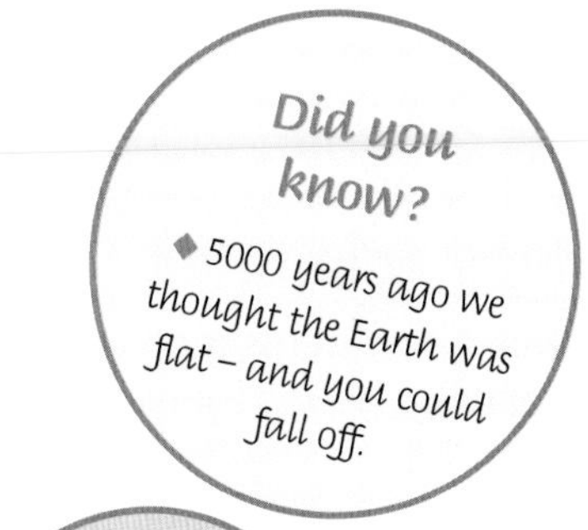

What if ...

... the Earth really was flat?

Your goals for this chapter

By the end of this chapter you should be able to answer these questions:

- In what ways am I connected to people and places all over the world?
- What are mental maps, and how can I improve mine?
- What does the scale on a map or plan tell me?
- What's the main difference between a sketch map, and the maps in an atlas?
- What are map grid references, and how do I use them to find places?
- How can I measure distance on a map?
- What are the compass points, and why are they useful?
- What are OS maps, and what kinds of things do they show?
- How is land height shown on an OS map? (Two ways!)

Did you know?

Maps over 4500 years old, drawn on clay tablets, were found in Iraq.

And then ...

When you finish the chapter, come back to this page and see if you have met your goals!

Your chapter starter

You are flying back to planet Earth to find Walter.

You have his address – but you don't want to ask for directions.

Would the images on page 14 help you to find him?

There are special drawings that would help you much more. Geographers just adore them. They're called ...... ?

Making connections

In this unit you'll see how we are connected to people and places all over the world – and how this can be shown using maps.

Walter connected

Walter. Alone in his room in Liverpool – but connected to people and places everywhere.

Mapping connections

Great Britain

The world

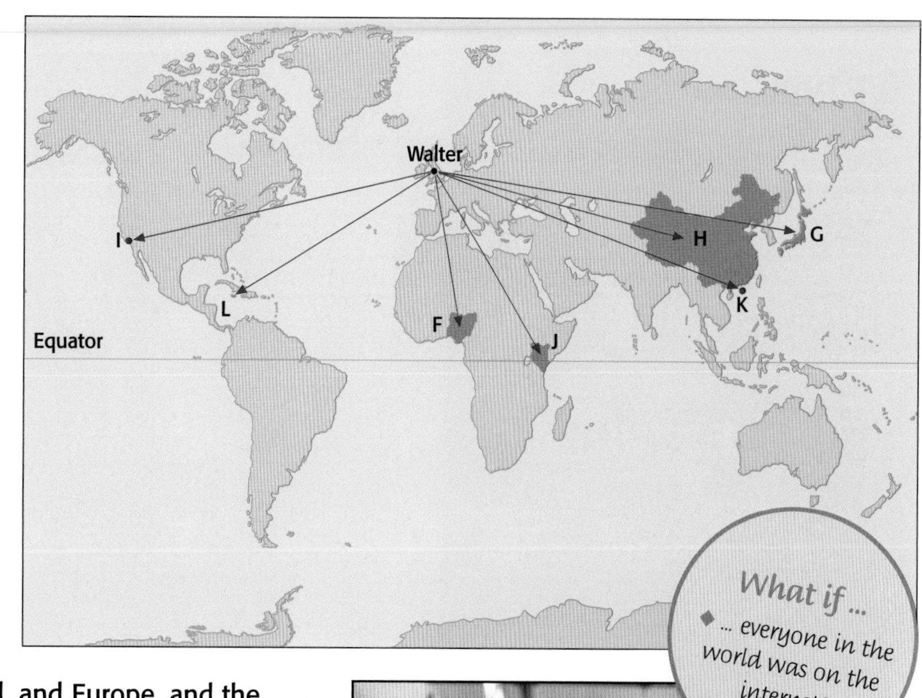

What if ...
- ... everyone in the world was on the internet?

Page 14 showed images of the world, and Europe, and the island where Walter lives (Great Britain, or just Britain). Above are maps of these places.

With maps it is easy to see where places are, and to show connections between them.

The maps above show Walter's connections from page 16. But that is just the start! All day long he is connected to *hundreds* of people and places – through school, TV, the internet, the things he owns or uses, the food he eats ...

It's just the same for you.

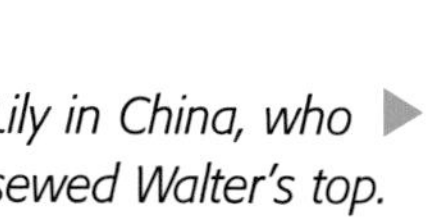

Lily in China, who sewed Walter's top.

Your turn

1 Match each letter on the maps above to a place named on page 16. Start like this: A =
(No peeking at the maps at the back of this book!)
Then give your answers to a partner to check.

2 Walter is connected to Jamaica by his CDs. That's an *international* connection. Pick out:
- a two other international connections for him
- b two local connections
- c two national connections

(Try the glossary?)

3 You too are connected to hundreds of places.
- a Make a big table like the one started on the right.
- b Leave room for three places, for each connection. Add more types of connection. (Music, TV, clothes?)
- c Now fill in the table, for you.

4 Imagine the UK is cut off from the rest of the world. No news, or letters, or TV, or phone calls, or food, or other goods, from other countries. And no internet.
- a List all the things *you* would have to do without.
- b What three things would you miss most?

Places I am connected to

Place	Connection
London	I've been there
... ...	Friends/relatives live there
... ...	I eat food that was grown there

A plan of Walter's room

In this unit you will learn what a plan is, and what a scale tells you.

A photo

This is Walter's room.
He tidied it for the photo.

A plan

This is a **plan** of Walter's room – a drawing of what you would see looking down from the ceiling.

A plan is really a map of a small area – for example a room, or a house, or your school.

The scale

The plan is much smaller than the actual room. In fact 1 cm on the plan stands for 30 cm in the room. That is the **scale** of the plan.

You can show scale in three ways:

1. In words: **1 cm to 30 cm**
2. As a ratio: **1 : 30** (say it as *1 to 30*)
3. As a line divided into cm, then labelled like this:

0 30 60 90 120 cm

The scale is always marked on a plan, so that people can tell the size in real life.

Plan of Walter's room Scale: 1 cm to 30 cm

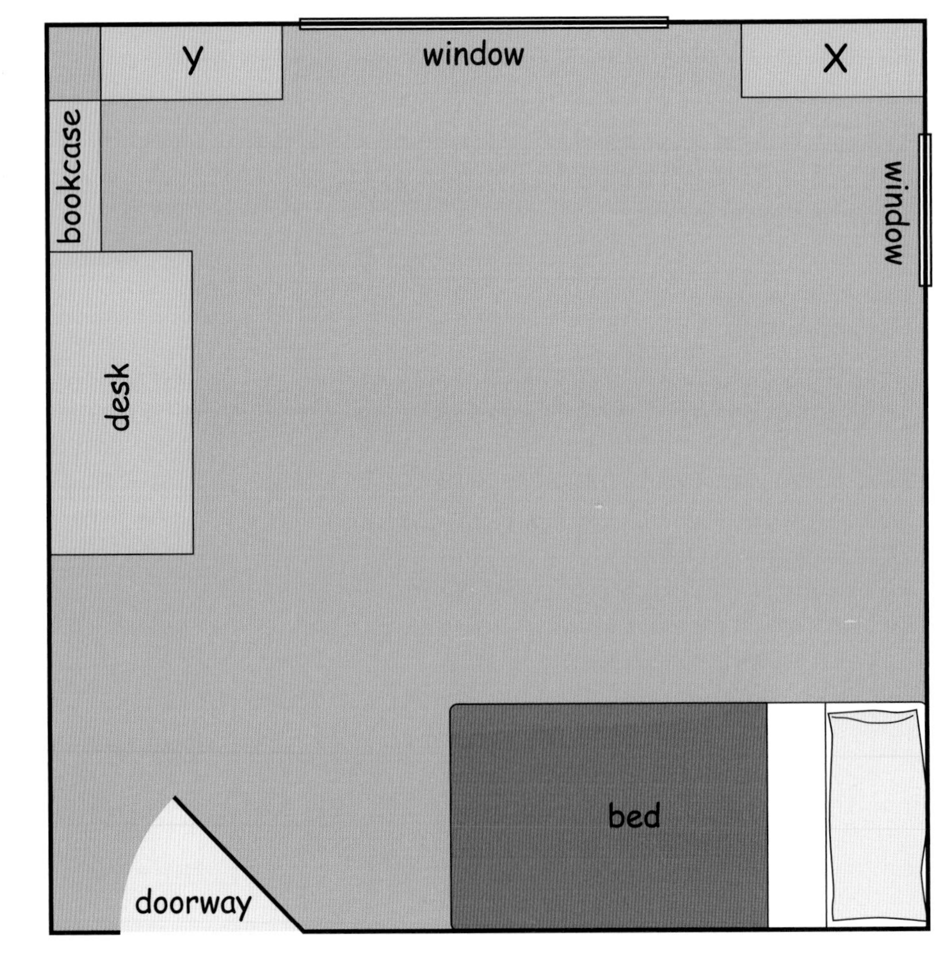

Working out scale

This is the plan of a table in Walter's kitchen. The table is 8 cm long in the plan. It is 160 cm long in real life.

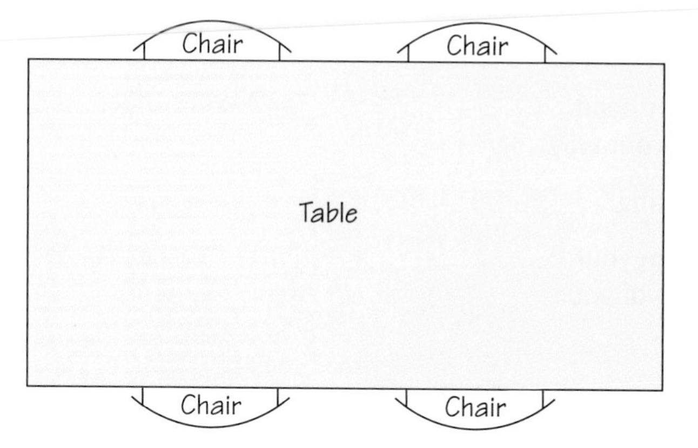

- 8 cm on the plan represents 160 cm in real life.
- So 1 cm on the plan represents 20 cm in real life.
- So you can write the scale as:
  1 : 20 or 1 cm to 20 cm or 0 20 40 cm

Be careful with units!

Look at this scale.

0 2 4 6 8 10 12 m

Here 1 cm represents 2 metres.
You can write this as **1 : 200.**

The 2 metres has been changed to centimetres. That's because *you must use the same units on each side of the* :

1 : 200 means **1 cm to 200 cm** or **1 cm to 2 m.**

Your turn

You will need a ruler for these questions.

1 Look at the plan of Walter's room. What do **X** and **Y** represent? (Check the photo!)

2 On a plan, one wall of a room is shown like this:

____

The scale of the plan is 1 cm to 60 cm.
How long is the wall in real life?

3 Below are walls from another plan. This time the scale is 1 : 50. How long is each wall in real life?

a ________________________

b ____________

4 Using a scale of 1 cm to 20 cm, draw a line to represent:
a 40 cm b 80 cm c 2 metres
Write the scale beside your lines.

5 If the scale is 1 : 300, what length does each line represent? Give your answer in metres.

a ______

b _________________________

c ______________

6 Draw a line to represent 1 kilometre, using each of these scales in turn:
1 cm to 1 km 1 cm to 50 m 1 cm to 100 m
Write the scale beside your line, in any form you wish.

7 Make a chart like this and fill it in for Walter's room.

Walter's room	On the plan	In real life
How wide is it? Measure the wall by the desk.		
How long?		
How long is the bed?		
How wide is the big window?		
How wide is the doorway?		

8 Walter is getting a new chest of drawers for his room:

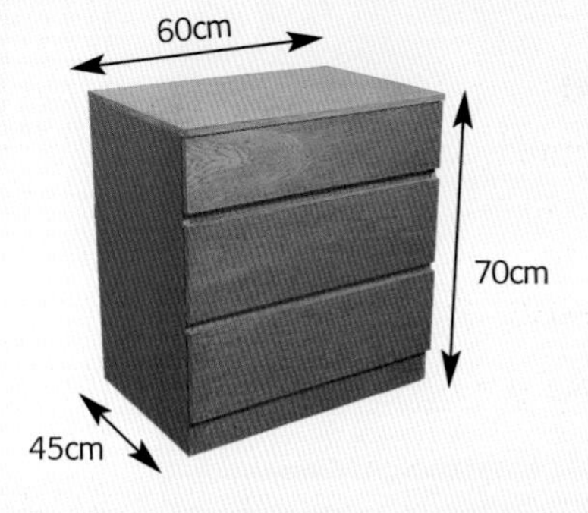

a To draw a plan of it, which surface will you use?
the top the side the front
b Draw the plan, to the same scale as Walter's room.
c Will the chest of drawers fit through the doorway?
d Where in the room would you put it?

9 Give three things in the photo of Arthur's room that are not on the plan. Suggest reasons why they're not.

10 How would you draw a plan for *your* room?
See if you can write a set of instructions to follow.

Your mental maps

You are a map maker! You have made lots of maps in your head.
Here you'll find out more about them – and get a chance to sketch one.

Mental maps

A mental map is a map that you make, and carry around, in your head. It is really a sequence of images, like a movie. It helps you find your way.

You have lots of mental maps. You use them without even thinking.

For example you have one of your home, that helps you get from your bedroom to the bathroom or kitchen, even in the dark. And one of your route from home to school, and to shops you like.

Sketching a mental map

It's fun to 'see' your mental map in your head, then draw it on paper. You end up with a rough map or **sketch map**.

Look at the sketch map on the right. Walter drew it from his mental map for his local area.

Do you think you'd find this map easy to follow?

▲ *Walter follows his mental map to the post office, with a parcel for Violet.*

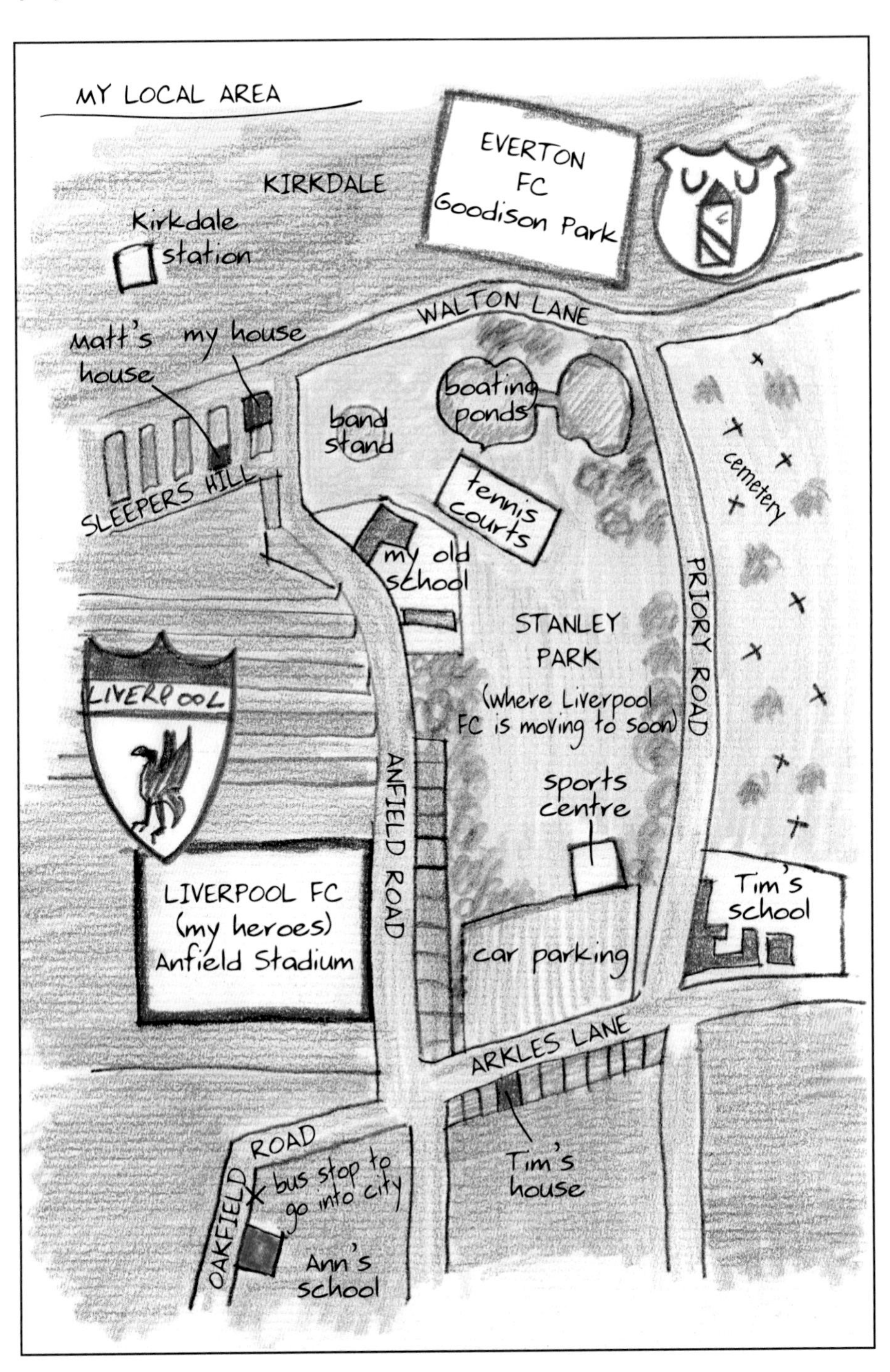

Your own mental maps

You have mental maps of your home, and your local area.

But that's not all. You have mental maps of other places you visit, and places you see on TV. You have mental maps of the whole UK, and even the world.

This shows Walter's sketch map of Britain, drawn from his mental map. What do you think of it?

They are gappy

Our mental maps show things that are important to us. Such as paths we use, shops we like, places we have fun.

But they leave out lots of things. Some have big big gaps. Some are quite wrong, and can get you lost.

You can make them better

You can make your mental maps better and better. The secret is: Look around. Keep your eyes open. Observe!

It's fun to build up your mental maps, and fill in places. It's like a game.

The better your mental maps are, the better your grasp of your world.

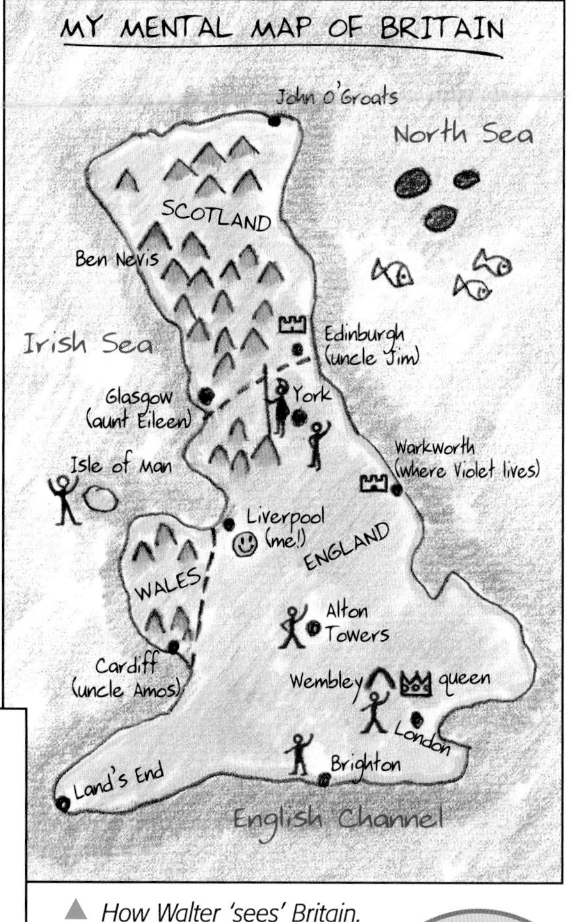

▲ *How Walter 'sees' Britain.*

What if ...
◆ ... your mental maps were nearly blank?

Your turn

1 What is a *mental map*?

2 Think about your mental maps. See how many you can list. For example, do you have one of your route from home to school?

3 Look at Walter's sketch map on page 20.
   a List the things he marked on it.
   b Beside each, say why you think he picked it.

4 Is Walter's sketch map easy to follow? Let's see! Give directions to get by road:
   a from Walter's front door, on Anfield Road, to Tim's house. You could start like this:
      - *Go out front door and turn right.*
      - *Walk along ____ ____ until ...*
      - *Then ...*
   b from Tim's house to the bus stop into town
   c from the corner of Walton Lane and Priory Road, to Anfield Stadium.

5 a Now, take a few minutes to picture the area around your school, in your head.
   b Using your mental map, draw a sketch map of the area. You can colour it in if you like.
   c Compare your sketch map with your partner's.
      i Do both show the same things?
      ii Do you think everyone's mental maps are different? Is that a good thing or a bad thing?

6 Look at Walter's sketch map of Britain, above. Compare it with the atlas map on page 139. Is the shape roughly right? Are his towns and cities in the right places? Give him a score out of 10.

7 Over the next week, pay special attention to the area around school. Look around. Keep your eyes open. Note the names of streets and roads. Observe! Then check: Is your mental map of the area changing?

2.4 Real maps

Here you'll compare a photo, a sketch map, and maps drawn to scale.

First, the photo

This photo shows Warkworth in Northumberland, where Walter's cousin Violet lives.

It's an **aerial photo** – taken from the air. Look at the loop of the river, and the Norman castle.

Next, the sketch map

Below is a sketch map of the same place, which Walter started. He drew it from the photo.

(You will do one later.)

Note that his sketch map has:

- a title, a frame, and a key
- some labels and annotations (notes)
- just enough detail to show the shape and layout of Warkworth. (Not each building and tree!)

▲ *Warkworth, from the air.*

B

Warkworth, where my cousin Violet lives. (Not to scale)

There is farmland all around the village.

These houses are tucked into the loop of the river.

bridges

castle

Key
- river
- trees
- homes and gardens
- road
- farm land
- open green areas

The castle was built over 900 years ago by Normans, but rebuilt later. Quite a lot of it is in ruins.

Now, a map drawn to scale

Look at this map of Warkworth.

It is not a sketch map. It is an accurate map, drawn to scale. See the scale below.

It uses symbols to show things. They are given in the key.

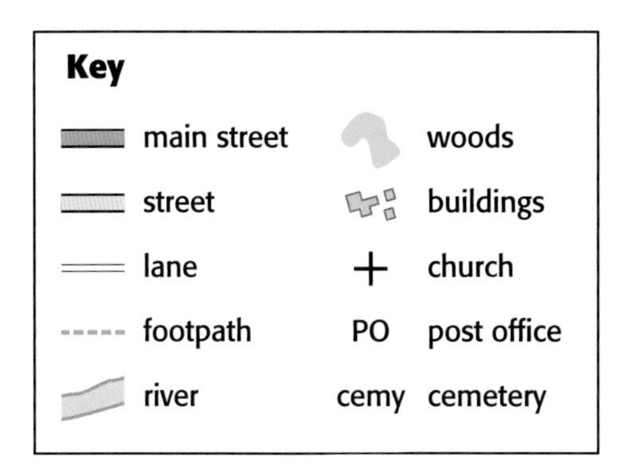

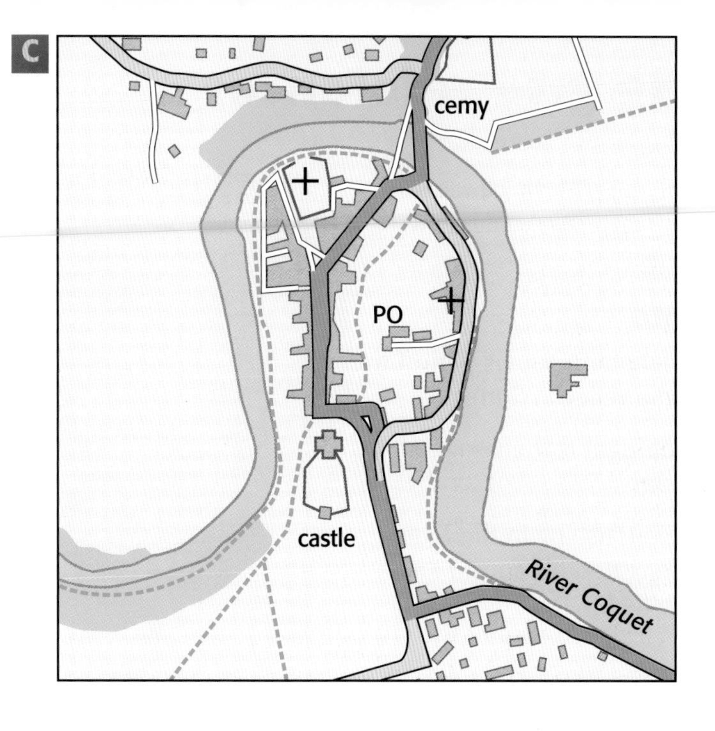

The same map with a grid

Here is the same map again. But this time, grid lines have been added.

They divide the map into squares. The columns and rows of squares have been labelled (A, B … and 1, 2 …).

Look at the post office. It's in square C3. The cemetery is in square C4.

(You always give the column first.)

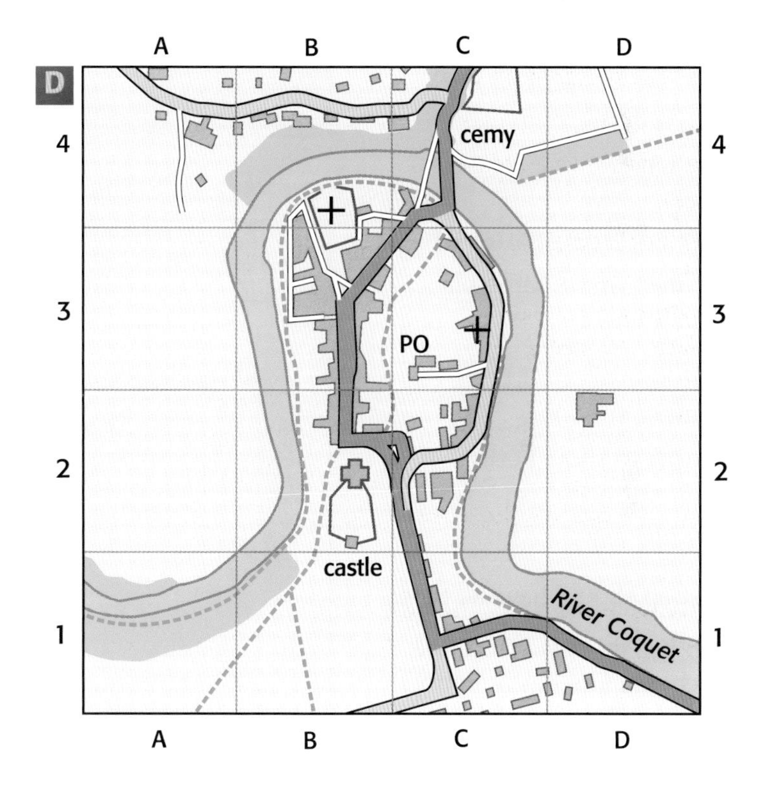

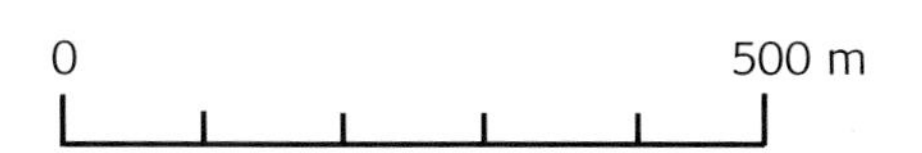

Your turn

1. Draw a sketch map (like the one Walter started) for photo **A** on page 22. Keep it simple. And don't forget:
   - a title, a frame, and a key
   - labels and annotations
2. Now swop sketch maps with your partner.
   - **a** See if you can agree on a fair way to score them. For example a mark out of 10 for the shape, 1 mark for each correct label, and so on. Write a list.
   - **b** Then give each other's maps a score.
3. Next, look at map **C** above. In which ways is it:
   a like your sketch map? **b** different from it?
4. Where is the castle, in map **C**? Tell us in words.
5. Look at map **D**, with its grid lines.
   - **a** In which square is: **i** the castle? **ii** the bridges?
   - **b** What is in: **i** square C3? **ii** square B4?
6. Do you think the grid lines in map **D** are a good idea, or a nuisance? Explain your answer.

Using grid references

In this unit you will learn how to find places on a map, using grid lines with numbers on.

A photo

This aerial photo shows part of the River Mole valley in Surrey.

In the top right is the village of Mickleham.

Walter went fishing in the Mole when he visited his cousin Kim. (The fish fled.)

A map of the same place

This is a map of the same place. Note that it has:

- a title
- a frame around it
- an arrow to show north
- a scale
- a key.

A good map should have all those.

The map has **grid lines** too. This time each grid line has a number (not like map D on page 23).

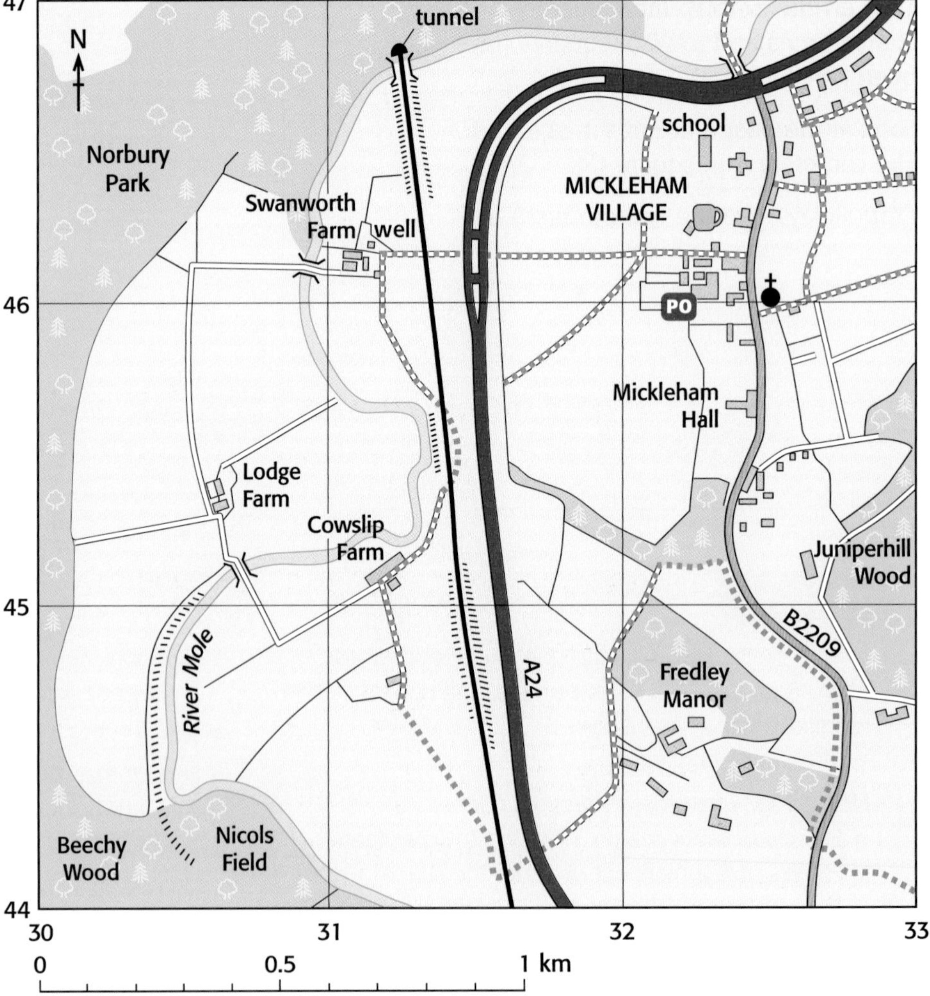

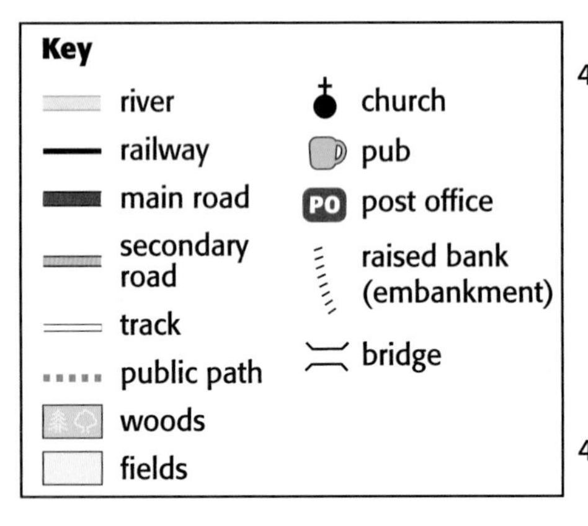

Four-figure grid references

The grid lines on a map help you find a place quickly.
To find the school in the square with **grid reference** 3246:

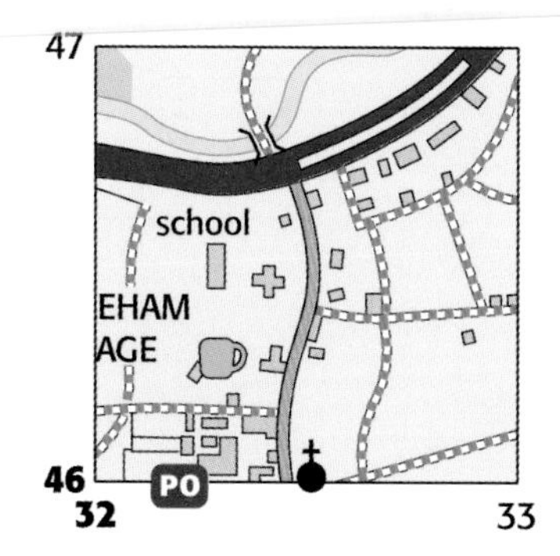

Find the square where lines 32 and 46 meet in the bottom left corner. (It's shown above.) Then look for the school.

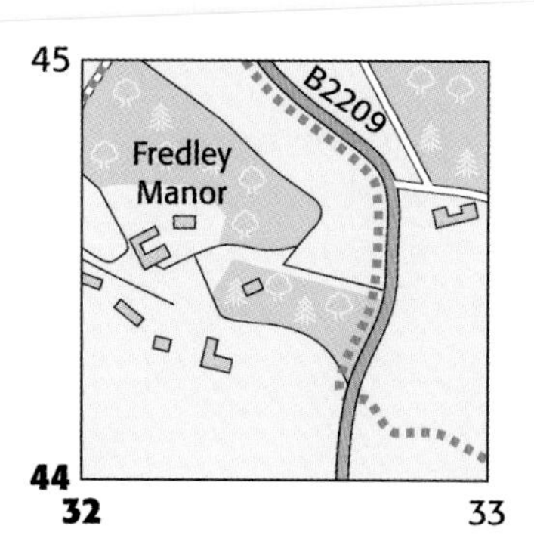

In the same way Fredley Manor is in the square with grid reference 3244. Lines 32 and 44 meet in the bottom left corner.

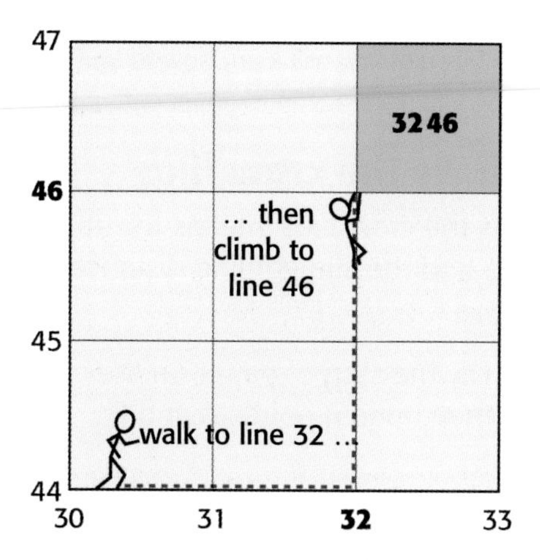

A grid reference always gives the number along the bottom first. This drawing shows how to find square 3246. *Walk before you climb!*

The grid references above are called **four-figure**. Why?

Six-figure grid references

There is a school *and* a church in square 3246 above.

You can say exactly where each is in the square using a six-figure number. Like this:

- Divide the sides of the square into ten parts, in your head, as shown on the right.
- Count how many parts you must walk along before you reach the building, and how many parts you must climb.

For the school you go 3 parts along and 5 parts up.
So its **six-figure grid reference** is 323465.
The one for the church is 325460. Do you agree?

And that's why grid lines with numbers are so good.
They let you give a position more exactly.

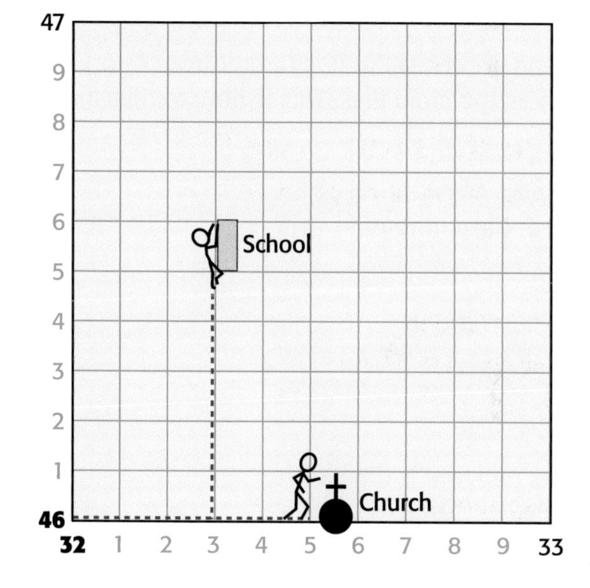

Your turn

1 Look back at the map on page 24. Give a four-figure grid reference for:
 a Mickleham Hall b Cowslip Farm c Nicols Field

2 What is at this grid reference on the map?
 a 312468 b 308448 c 309461

3 Give a six-figure grid reference for:
 a Mickleham Hall b the post office c the pub

4 There is something at 312463 that you can't see on the photo. What is it?

5 You can't see the river on the photo. How can you tell where it is?

6 Describe what you will see, if you stand at 313453 facing south. (With your back to the north!)

7 How far is it from Lodge Farm to Cowslip Farm, along the track? (Think of a way to measure it using the scale.)

8 This shows a signpost in the area. Where do you think it belongs on the map?
Write a six-figure grid reference for it.

How far?

In this unit you will learn how to find the distance between two places on a map.
You will need a strip of paper with a straight edge.

1 As the crow flies

'As the crow flies' means the straight line distance between two places.
To find the straight line distance from A to F, this is what to do:

Key
—— road

1. Lay the strip of paper on the map, to join points A and F.
2. Mark it at A and F.
3. Now lay the paper along the scale line to find the distance AF.

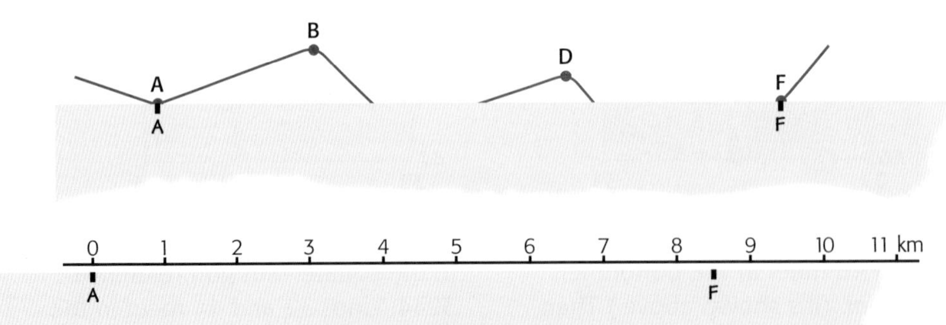

From A to F as the crow flies is 8.5 km

2 By road

Roads bend and twist. So it is further from A to F by road than as the crow flies. This is how to measure it:

1. Lay the strip of paper along the straight section of road from A to B.
2. Mark it at A and B.
3. Pivot the paper at B until it lies along the next straight section, B to C. Mark it at C.
4. Now pivot it at C so that it lies along the next straight section, C to D. Mark it at D.
5. Move along the road in this way, section by section, until you reach F.
6. Place the paper along the scale line to find the distance AF.

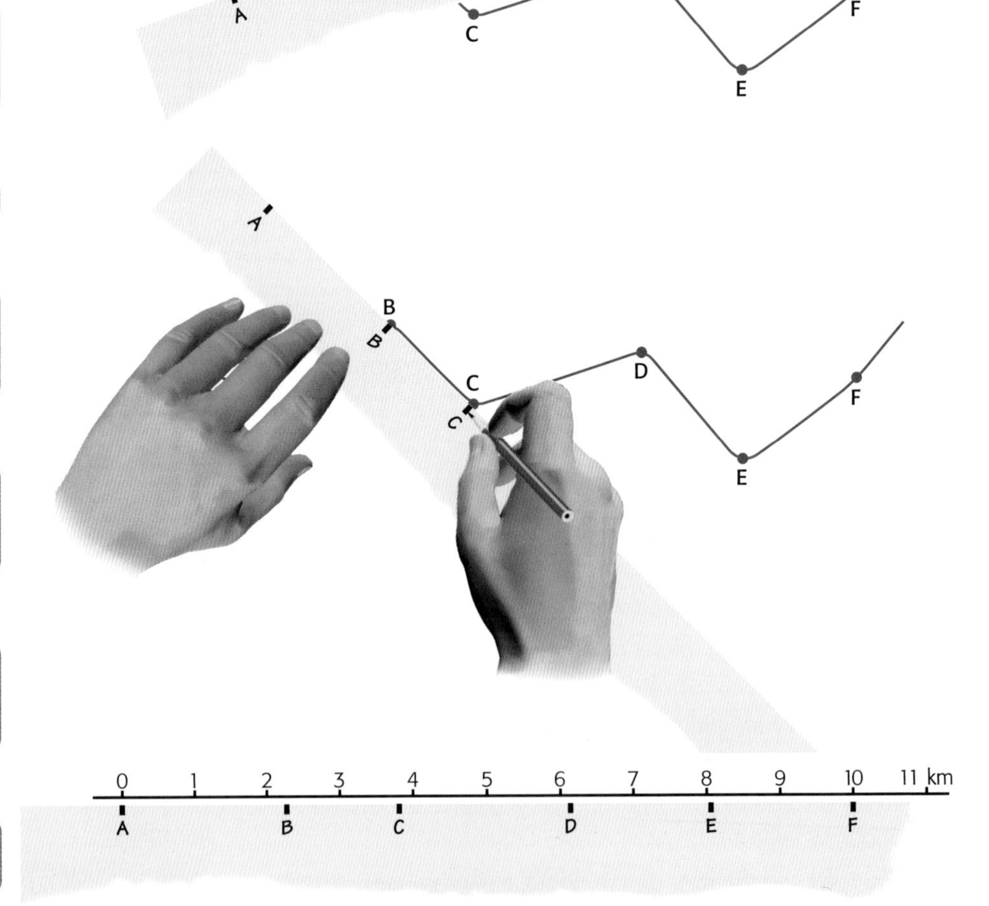

From A to F by road is 10 km

Your turn

The photo and map on page 24 showed part of the River Mole valley in Surrey. This map shows more of the same area. (Are both maps at the same scale?)

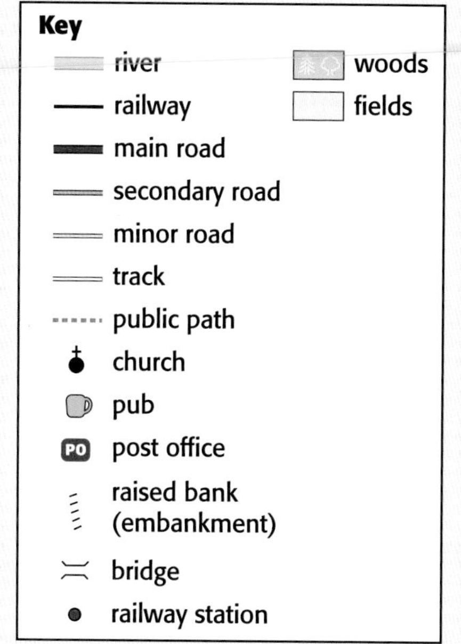

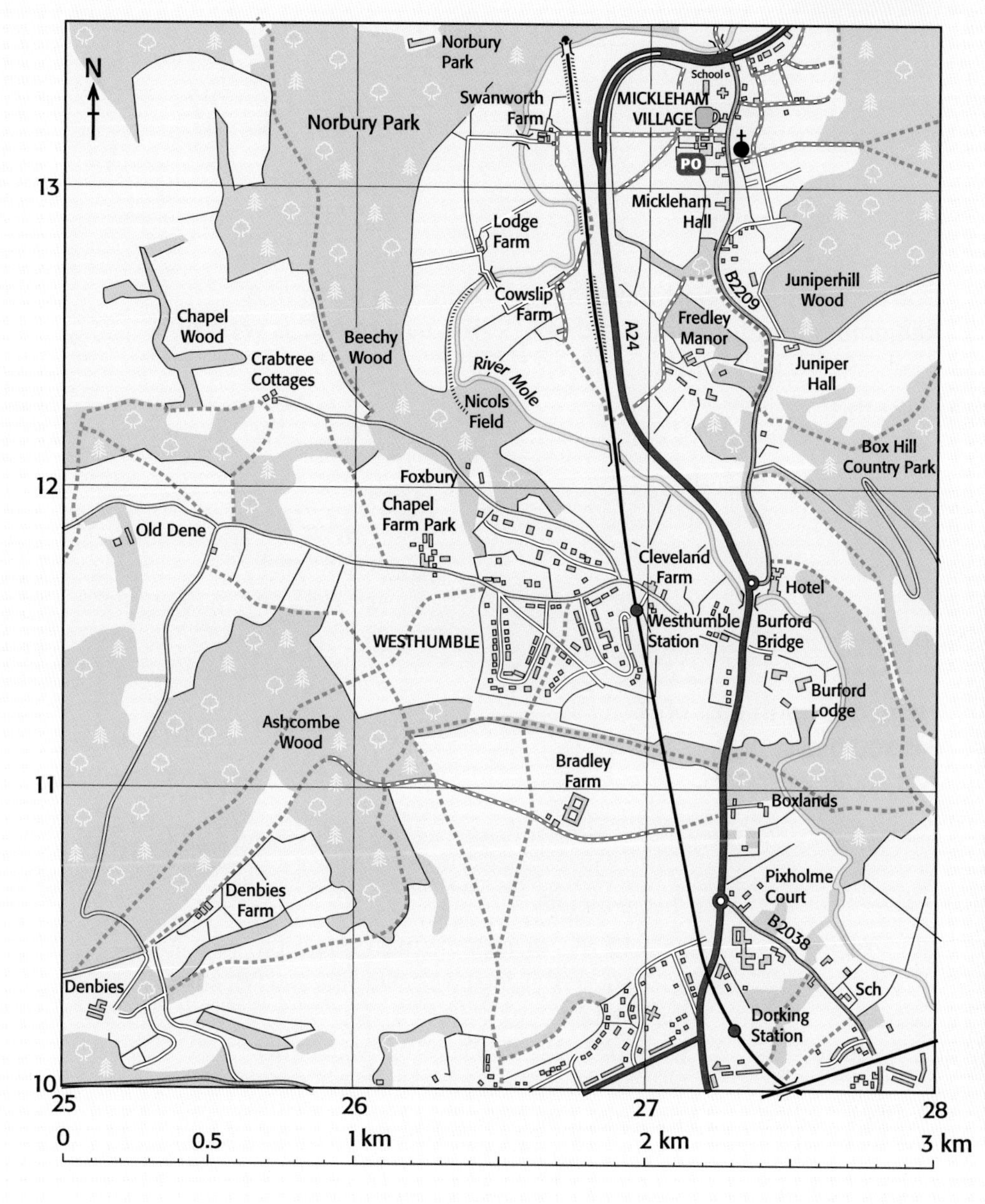

▲ *Boxlands.*

▲ *Juniper Hall.*

1 How far is it as the crow flies from Mickleham church to Westhumble station?

2 How far is it by rail from Westhumble station to Dorking station?

3 About how far is it by road from Mickleham Hall (273129) to the hotel at 274117?

4 Walter hired a bike at Westhumble station. He followed these directions:
Go along the short road from the station to the T-junction at Cleveland Farm. Turn left. At the next fork, take the road to the left and cycle for 0.7 km.
Where did he end up?

5 Every day, Kim's mother collects her from school at 276103 and drives her home by this route:
From the school, go right on the B2038.
At the roundabout, take the A24 north for 0.9 km.
Turn left onto the minor road and continue for 0.5 km.
Now take the road to the right and continue for 1.4 km.
Where does Kim live?

6 Juniper Hall and Boxlands are shown above.
  a Find them on the map, and give six-figure grid references for them.
  b Write instructions telling a friend how to get from Juniper Hall to Boxlands. Don't forget distances!

2.7 Which direction?

In this unit you will learn how to give and follow directions, using N, S, E and W.

The compass points

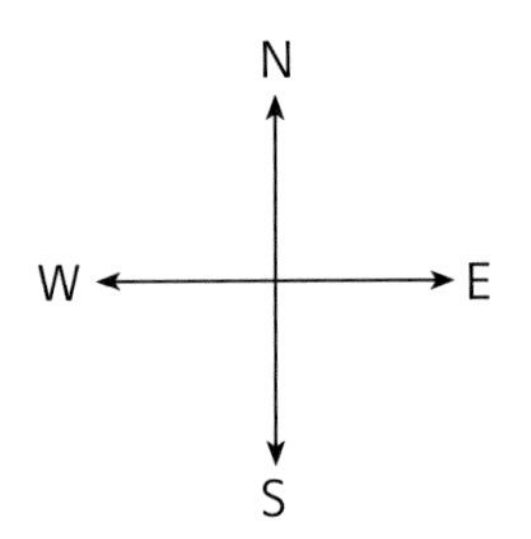

N, S, E, W are the four compass points: north, south, east, west.

Don't get east and west mixed up. Remember they form the word ***we***.

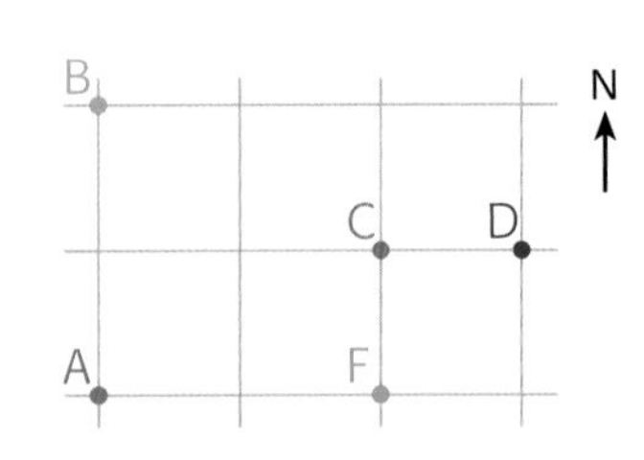

Here B is north of A. F is east of A. C is west of D.

We can add other directions in between, like this:

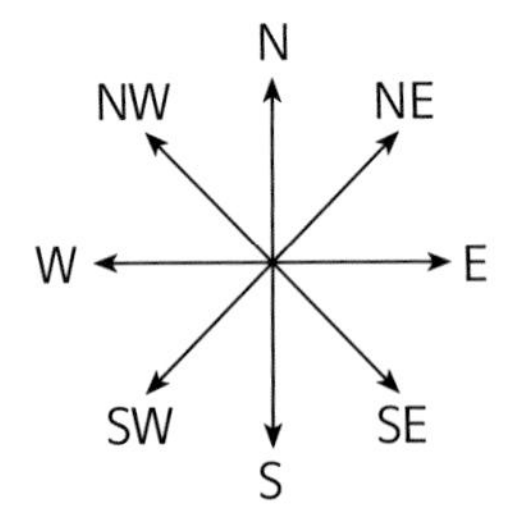

NE stands for north east (or north *of* east). SW stands for south west (or south *of* west).

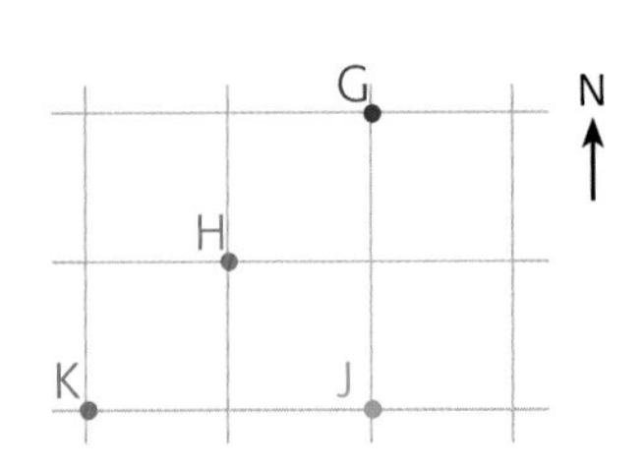

Here, G is north east of H. J is south east of H. K is south west of H.

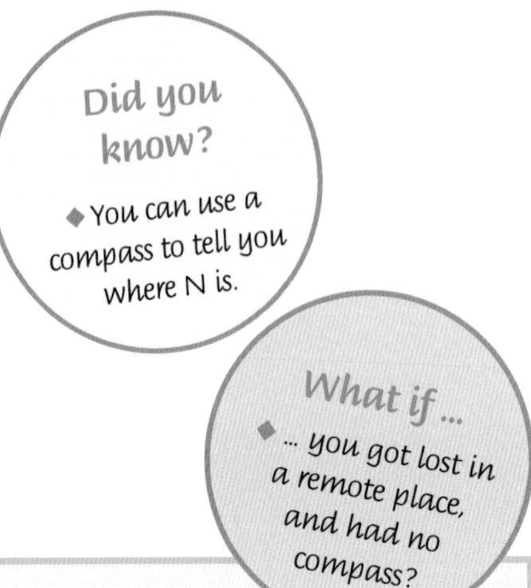

Your turn

1 You are standing at C in the first grid above. Which direction do you face when you turn towards:
a F? **b** D? **c** A? **d** B?

2 Page 29 shows where Walter went on holiday. The bowling alley is in square D5. What is in square:
a A10? **b** F6? **c** C4? **d** F2?

3 You are at the hostel. In which direction is :
a bike hire?
b the riding school?

4 In which direction would you go, to get to:
a the duck pond, from the pizza place?
b the gym, from the bowling alley?
c bike hire, from the kite shop?

5 How far is it by footpath from the door of the hostel to the door of the bike hire shop? Use your ruler, and then the scale.

6 To get from the cafe to where Walter stayed:
- From the cafe door, walk 50 m SE, then 65 m N.
- Next walk 40 m E, then 10 m SE, then 10 m SW.

Where did he stay?

Treasure hunt

7 Look for the ● near the main gate. From here, if you go 2 squares N, then 1 square NW, you will arrive at the letter ⓐ.
Now follow the directions below, in order.
For each, write down the letter you arrive at.
The letters will make a word.
- Start at ●. Go 2 squares W.
- Then go 8 squares N and 4 squares E.
- Then go 1 square N and 5 squares W.
- Next, go 2 squares SE then 4 squares S.
- Then go 2 squares SW and 1 square SE.
- Then 3 squares NW, followed by 4 squares E, then 3 squares NE, then 2 squares N.

What word have you made?

8 a Now choose your own word, with at least 5 letters but not more than 8.
b Write instructions for making this word, like those in question 7. Start from the ●.
c Ask a partner to follow the instructions.

Map of your holiday village

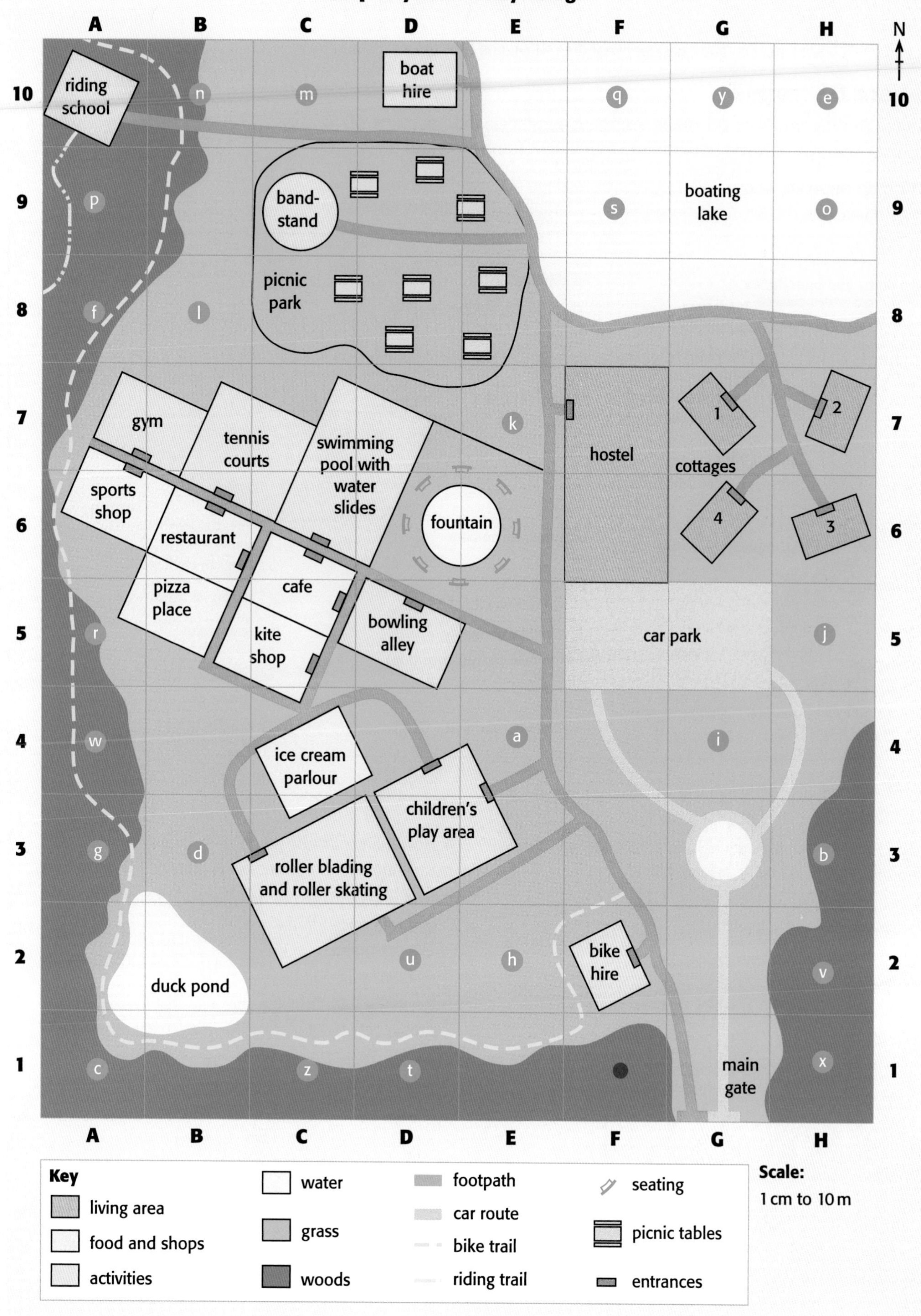

Key
- living area
- food and shops
- activities
- water
- grass
- woods
- footpath
- car route
- bike trail
- riding trail
- seating
- picnic tables
- entrances

Scale:
1 cm to 10 m

2.8 Ordnance Survey maps

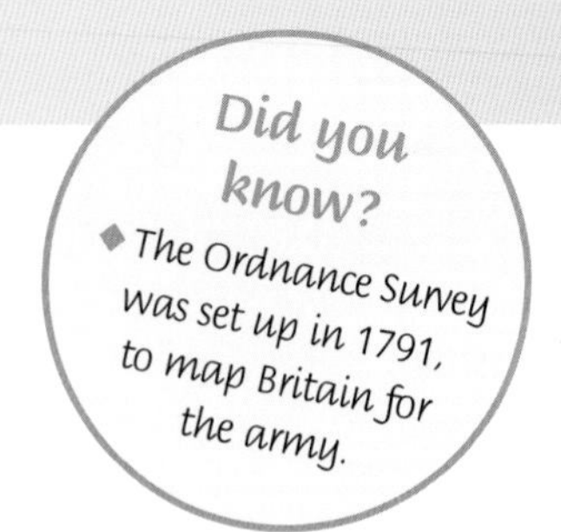

In this unit you'll learn what OS maps are, and what they show, and how to use them.

What are OS maps?

Ordnance Survey maps or **OS maps** are maps of places, that give lots of detail. They use symbols to show things. They have numbered grid lines.

The OS map opposite shows Warkworth (from page 22), and Amble. The key below has the symbols. And there's a larger key on page 138.

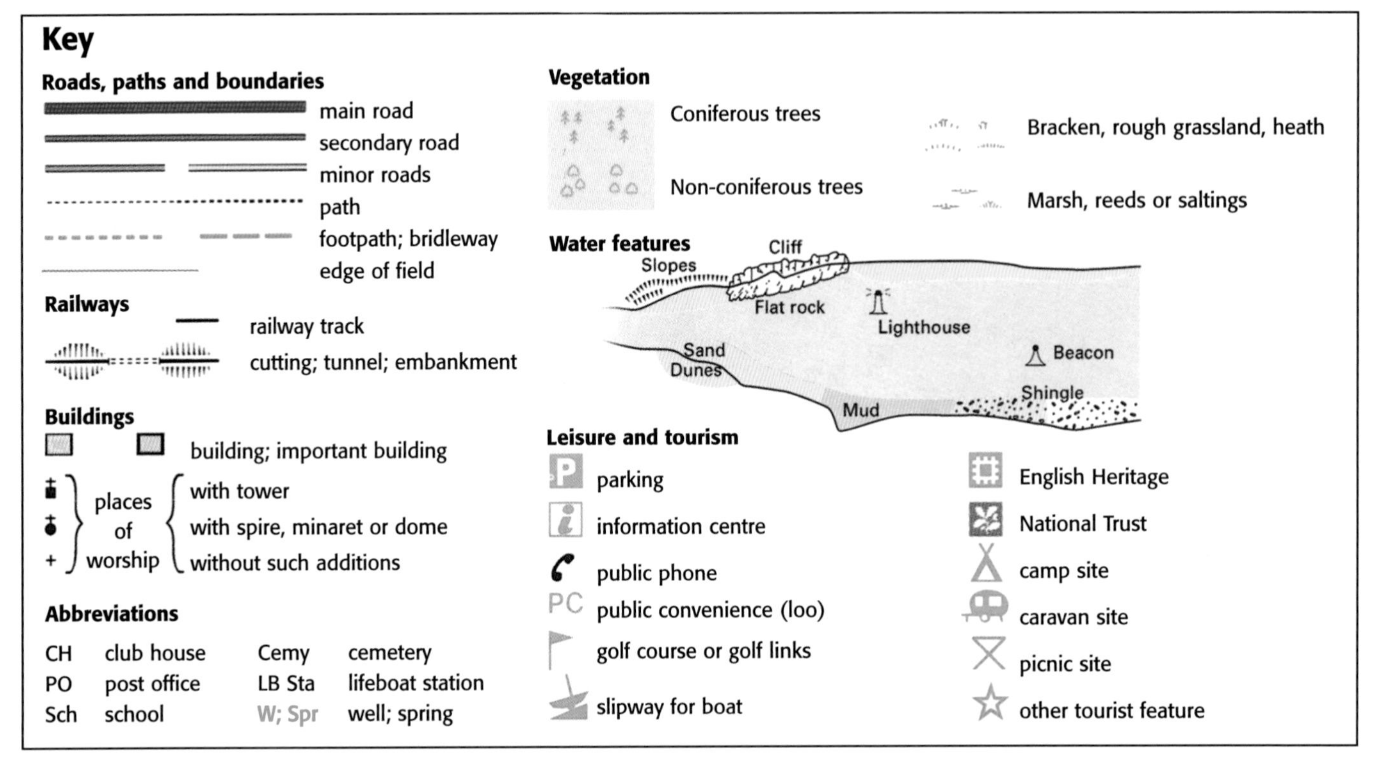

Your turn

1 Look at Warkworth on the OS map. Name the river that flows through it. Where is it flowing to?

2 Find it on the map, and give its four-figure grid reference:
a Northfield   b Gloster Hill   c North Pier

3 What is at this grid reference, on the map?
a 243045   b 275041   c 247057
d 243065   e 249063   f 273051

4 4 cm on this OS map represents ____ in real life?

5 The top of an OS map is always north. Look at the photo of Warkworth on page 22. Where is north on it?

6 Violet's house is marked on the photo on page 22. Find it. Then find it on the OS map, and see if you can write a six-figure grid reference for it.

7 Warkworth has a population of 1600. Now look at Amble. Its population is one of these. Which one?
a 1000   b 2000   c 5600   d 9300
How did you decide?

8 How many of these are there in Amble?
a schools   b places of worship   c cemeteries

9 Find one of these on the map and give a six-figure grid reference for it:
a a post office   b a club house
c a public phone box   d a mast

10 What clues are there on the map that Warkworth and Amble get lots of visitors? Give as many as you can.

11 What is there for tourists to do, around Warkworth and Amble? Using the information on the map, write a list.

12 On the map, what clues are there that the coast and sea around Amble might be dangerous?

13 Violet used to go to the school in square 2503.
a How far is it from her home? (Use the scale.)
b Pretend you are Violet. Draw a sketch map of your route to that school. Mark in what you think are the key things you'd see on the way.

OS map showing Warkworth and Amble

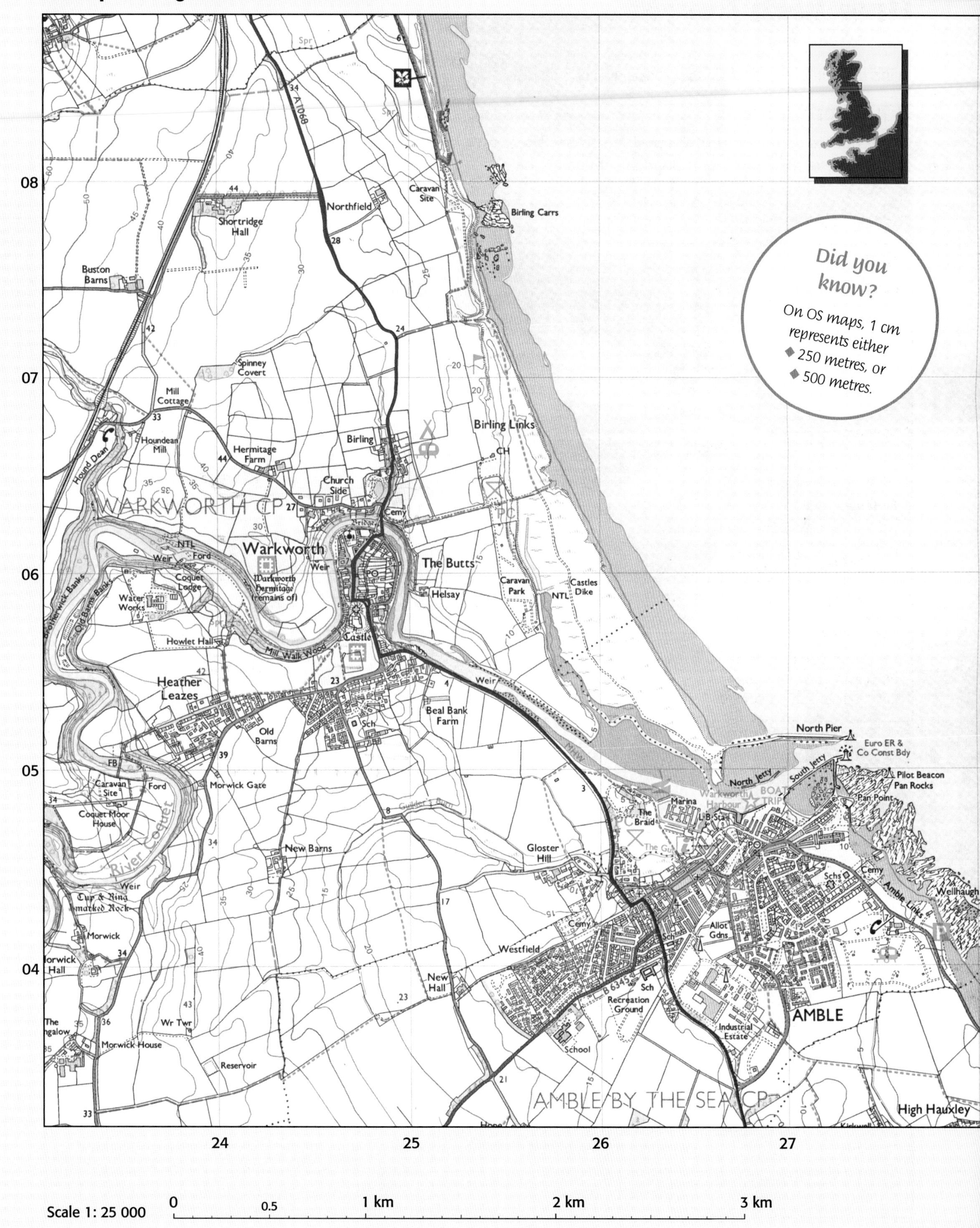

How high?

In this unit you'll learn how height is shown on an OS map.

A hilly problem

These photos show Alton Towers, where Walter spent time shrieking. So what's the land like around here? Is it flat? Or hilly?

▲ *Better before lunch?*

The OS map below shows the area around Alton Towers – and tells you how flat or hilly it is. The map shows height in two ways …

1 Contour lines. Everywhere along a contour line is the same height above sea level. The number on the line shows the height in metres. Here, the lines are every 10 m above sea level.

2 Spot heights. They give the exact height at a spot, in metres above sea level.

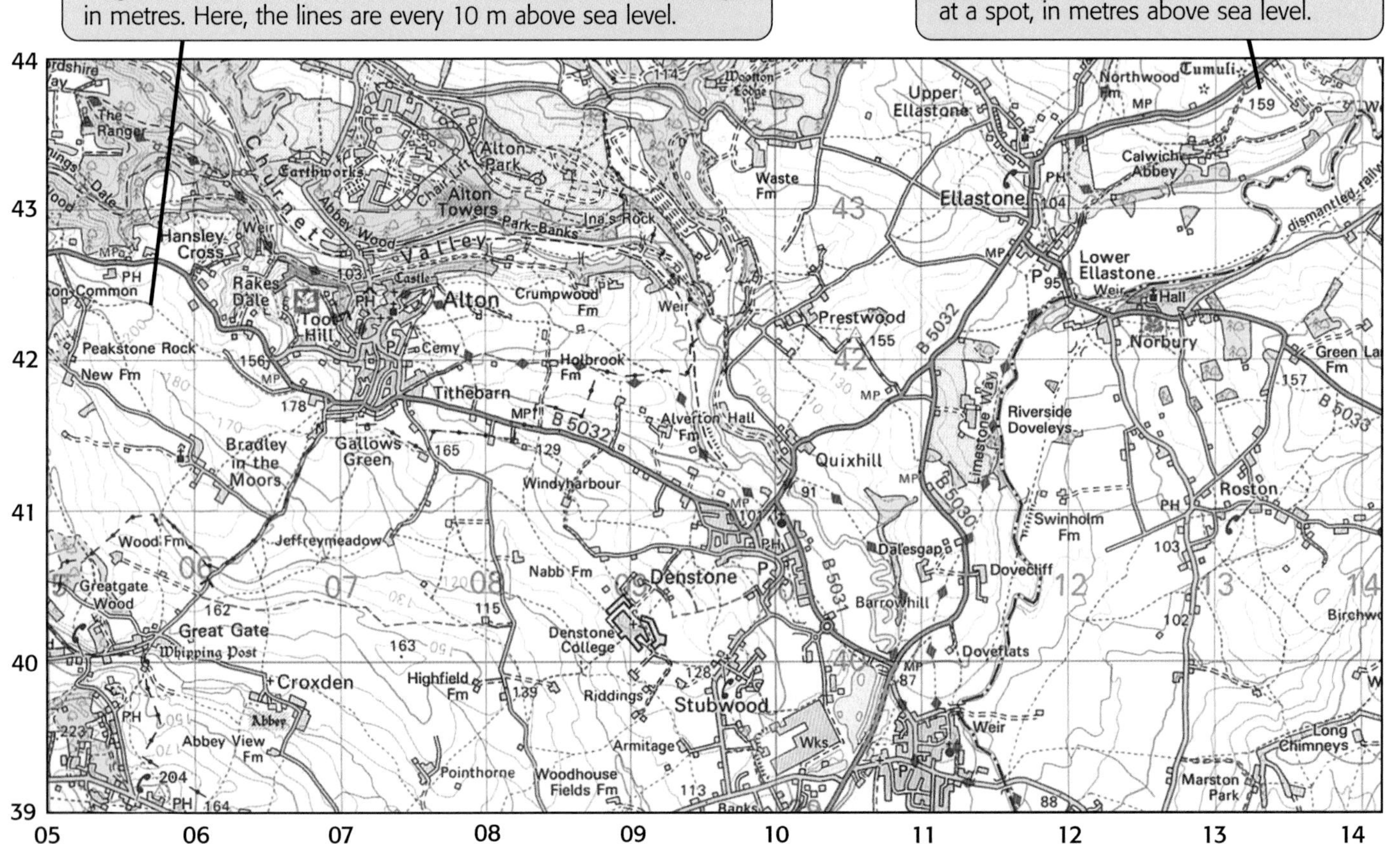

More about contour lines

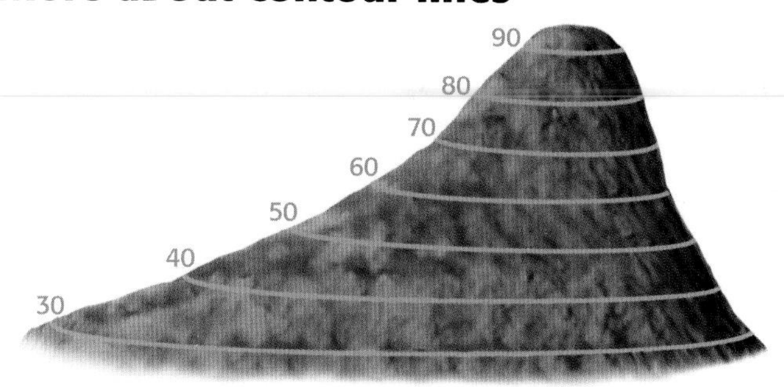

The contour lines are marked on this hill at 10 metre intervals. On a map, you see them from above …

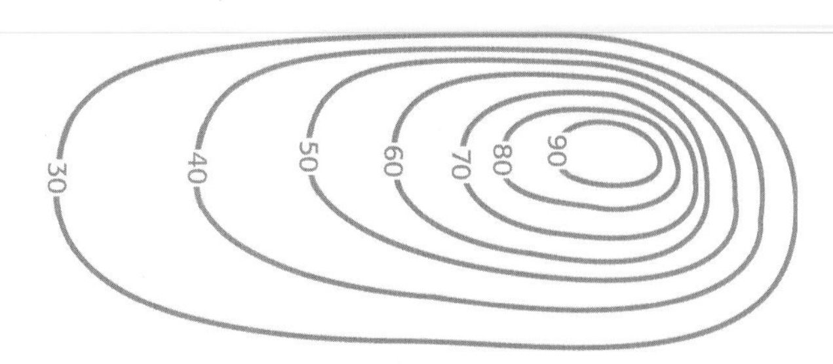

… like this. They are close together where the slope is steep, and further apart where it is gentle.

Remember:
- where contour lines are very far apart, it means the ground is flat.
- where they are very close together, the ground slopes steeply.

Did you know?
- Some places in the UK are below sea level. (See page 139.)

Your turn

1 Match the drawings to the contour lines.
Start your answer like this: A =

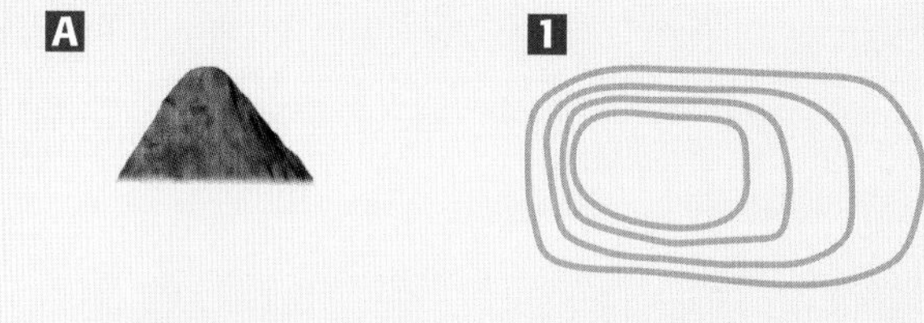

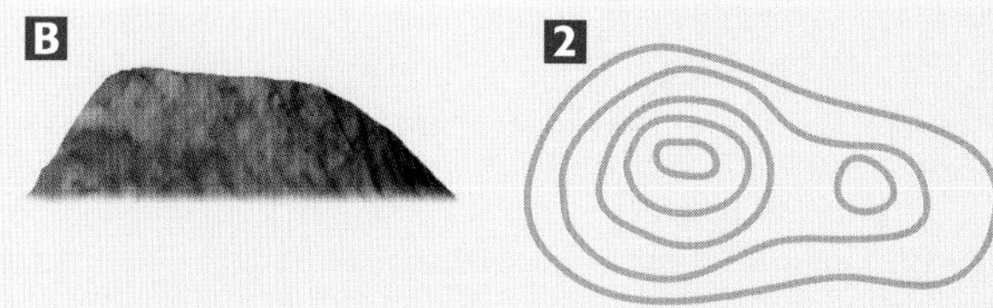

These questions are about the map on page 32.

2 a In which square is the main part of Alton Towers?
b Why did they call it Alton Towers?

3 Look at the pointer to the left of Alton Towers.
a It points to a line marked 200. What does the number tell you?
b Look at the next line down. What does it represent?

4 About how high above sea level is:
a Quixhill (1041)?
b Highfield Farm (0739)?
c the car park (P) just south of Ellastone (1142)?

5 a Which square on the map do you think has the steepest land? How can you tell?
b Which square has the most flat land? Did you have any problem in deciding?

6 What can you say about the land around the Alton Towers theme park? Is it flat, or a bit hilly? Explain.

7 Going by road, say whether it is uphill, downhill or along flat land:
a from the phone box at 056391 to the phone box at 052400
b from the bridge at Quixhill (100412) to Prestwood (103423)
c from the roundabout at 103403 to the church at 100409

8 Look at the River Churnet, flowing past Alton.
a Is it flowing along steep land, or flat land? Give your evidence.
b Which way is it flowing: towards the top of the map, or the bottom? How did you decide?

9 You have to plan a *long* walk for a group of students, starting and ending at Waste Farm (0943).
- It must be *at least* 10 km long.
- You must keep to roads, tracks and footpaths.
- It must have variety!
- Include something historical if you can.

a Plan your route and draw a sketch map of it. Mark in interesting things you will see or visit along the way.
b Work out the length of your walk, from the OS map.
c Give your sketch map a title and a north arrow, and say whether it is to scale or not.

Settlement

The big picture

This chapter is about settlements – the villages and towns and cities we live in. These are the big ideas behind the chapter:

- For nearly 200 000 years, we humans moved around in search of our food: berries, seeds, and animals to kill.
- Then, about 10 000 years ago, we discovered how to farm – and we began to settle down.
- When we found a good place to settle, we built shelters. Over time, the clusters of shelters grew into villages, and towns, and cities.
- Today, many of our settlements are still growing – and we are still building shelters. (We call them houses!)

Did you know?

- *The settlement we call London is less than 2000 years old.*
- *Damascus, in Syria, is over 10 000 years old.*

Your goals for this chapter

By the end of this chapter you should be able to answer these questions:

- What did our ancestors think about, in choosing a place to settle in?
- What do these terms mean?

  settler *settlement* *site*

- What kinds of things cause a settlement to grow?
- What are these? Where in a settlement am I likely to find them? Why?

  the CBD *terraced housing* *modern housing*
  old industrial area *modern industrial area*

- What sort of clues does an OS map give me, about land use in a town or city? (At least four examples.)
- Why does the UK need more homes?
- What do these terms mean?

  developer *development* *redevelopment*
  greenfield *brownfield* *commute* *dormitory town*

- What does *sustainable development* mean? What are its three strands?
- What kinds of questions can I ask, to see if a development is sustainable?

Did you know?

- *24 of the world's cities have over 10 million people each!*

What if ...

- ... everyone moved into cities?

What if ...

- ... we all lived in tents and moved around?

And then ...

When you finish the chapter, come back to this page and see if you have met your goals!

Your chapter starter

Page 34 shows a settlement.

What's a settlement?

Pick out five key things you notice about this one.

Do you think it has always looked like this? Give your reasons.

In what ways is your settlement like this one? In what ways is it different?

Settling down

In this unit you'll find out what we humans looked for, when choosing a place to settle in.

Once upon a time

As you saw on page 10, Earth was empty for billions of years. Then life began. Around 200 000 years ago …

… the first humans like us appeared. They lived by eating fruit and berries, and hunting …

… which meant they were nearly always on the move, chasing dinner. Then one day …

… around 10 000 years ago, they noticed something amazing: if you drop seeds in soil, plants grow!

So they began to settle down in one place and grow their food. These were the first farmers.

They chose places or **sites** to settle in, that had what they needed. Like good flat land, water, firewood…

… shelter from wind and rain … material for making things (wood, stone, clay, and later, metals) …

…protection from enemies … and easy access to other places, for trading.

They cleared the land and planted crops and put up dwellings. The result – a **settlement**.

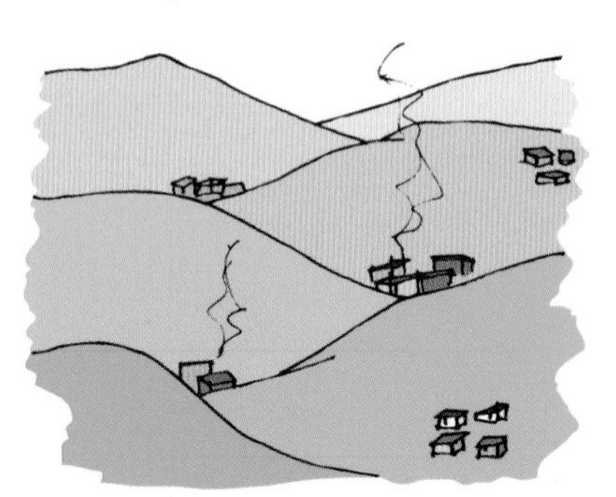

Years passed. The numbers of humans – and settlements – grew.

Some settlements grew larger and larger. And now …

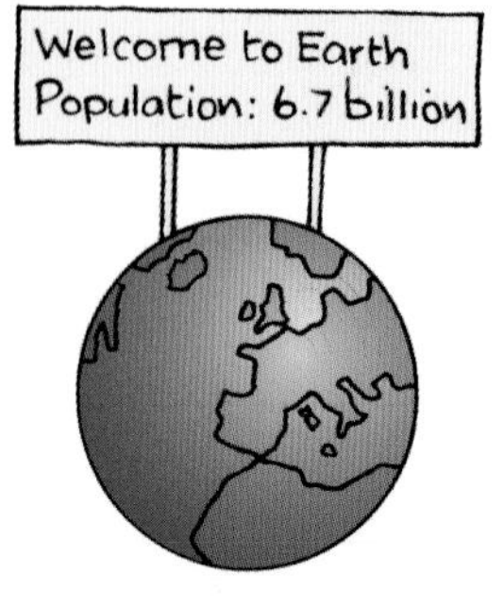

… there are more than 6.6 billion humans, and half of us live in cities.

Your turn

1 It is 5000 BC. You are leading your tribe on a search for a place to settle in. Draw a spider map showing the factors you will consider, when choosing a site – like this:

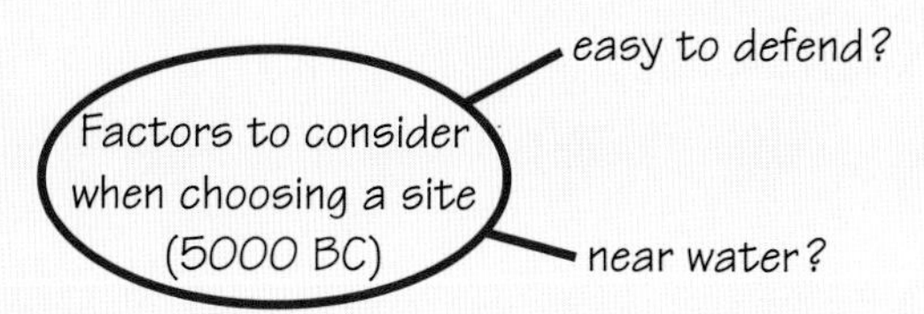

2 Copy and complete in your own words:
   a A settlement is …
   b A site is …
   c The early settlements were usually situated near …
3 Look at the photos A to E. For each photo, decide whether it shows a settlement or not. Give your reasons.
4 For each *settlement* in the photos, suggest reasons for choosing that site. (Try to give at least two reasons for each.)
5 The government wants to build a new town in the UK, starting next year.
   a What factors do you think will be important, when choosing a site for it? List them, and say why they are important. (What about good road links? And a water supply?)
   b Underline any factors that did *not* apply in 5000 BC.
6 Now think about *your* settlement. Why do you think people first chose that spot to settle in?

A in the Philippines

B in Canada

C in Morocco

D in France

E in Switzerland

3.2 Example: settling in Aylesbury

Here you'll find out who settled first in Aylesbury – and why.

Once upon a time

7000 years ago, Britain was covered in thick forest. Only small groups of hunters lived here.

Then, around 6000 years ago, farmers arrived from mainland Europe. They brought animals, and seeds, and ploughs. They cleared land, and settled down, and started to farm.

As time passed many other groups arrived too. Look at this table.

The new arrivals

Who?	Around when?
Early farmers	4000 BC
Celts	800 BC
Romans	40 AD
Saxons	500 AD
Vikings	850 AD
Normans	1066 AD

Who settled in Aylesbury?

Aylesbury is a town about 55 km from London. It's in the middle of an area of good flat farmland.

We know the Celts spent time here, and the Romans built a road through the area, from London.

But the first people to really settle down here were the Saxons.

This sketch map shows what they may have found when they arrived, in 571 AD, over 1400 years ago.

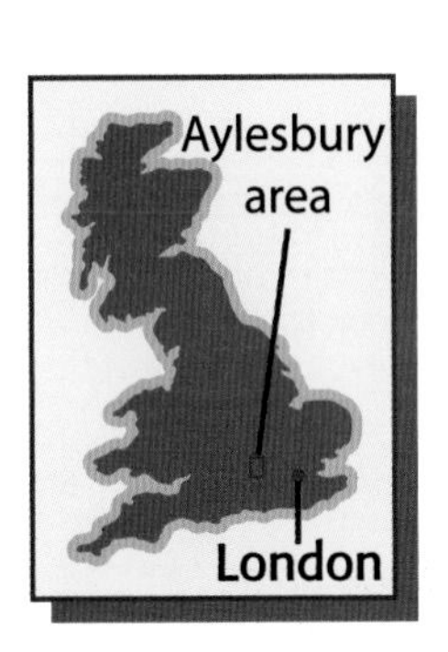

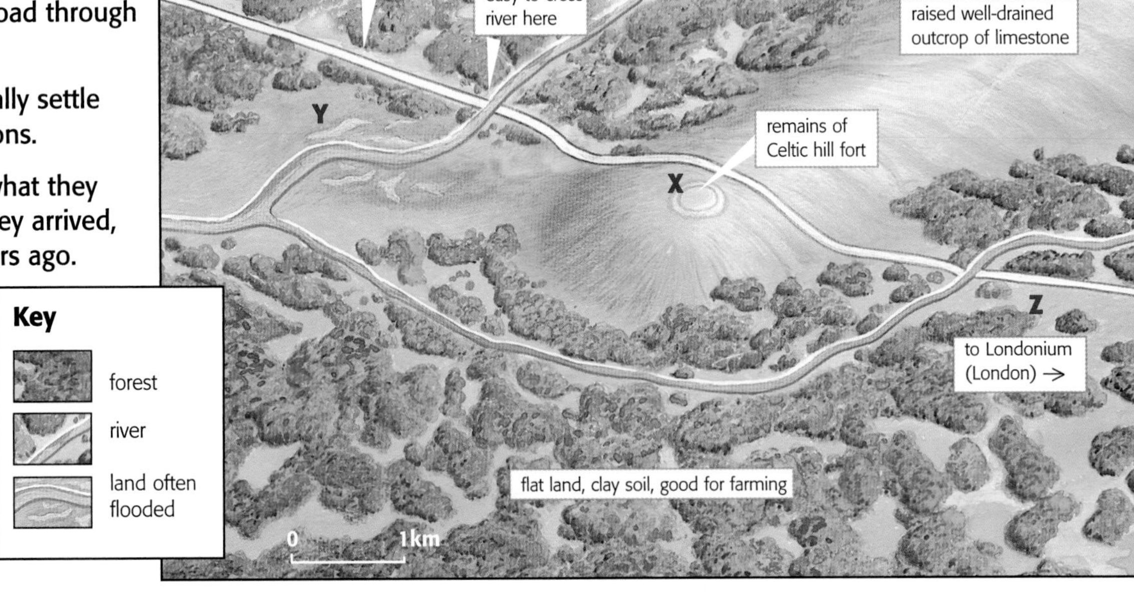

The settlement starts ...

The Saxons liked what they saw, and decided to stay.

Perhaps there were only a few families, to start with. The first thing they did was build huts to live in. And then they began to farm.

When news got around that the farming was good, more families turned up. So more huts were built. The settlement grew.

As time passed, the Saxons became Christian, and built themselves a church.

What a Saxon settlement looked like – we think! ▶

... and grows, and grows

The little Saxon settlement kept on growing. Compare this sketch map and this photo.

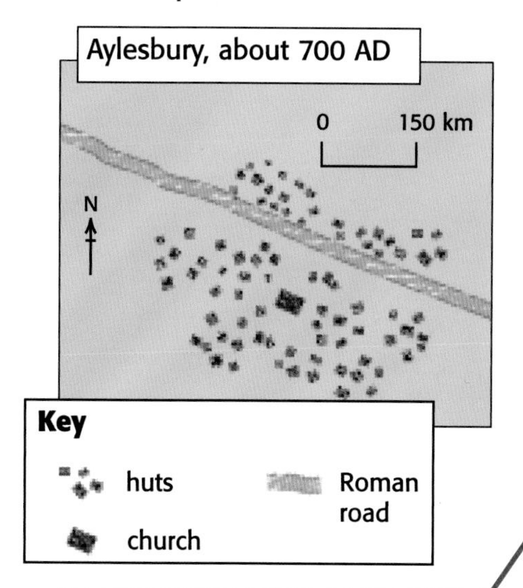

The Roman road still runs through Aylesbury. Now it's called the A41.

This is a 13th century church. It was built on the site of an early Saxon church ... which was built on the site of the Celtic hill fort.

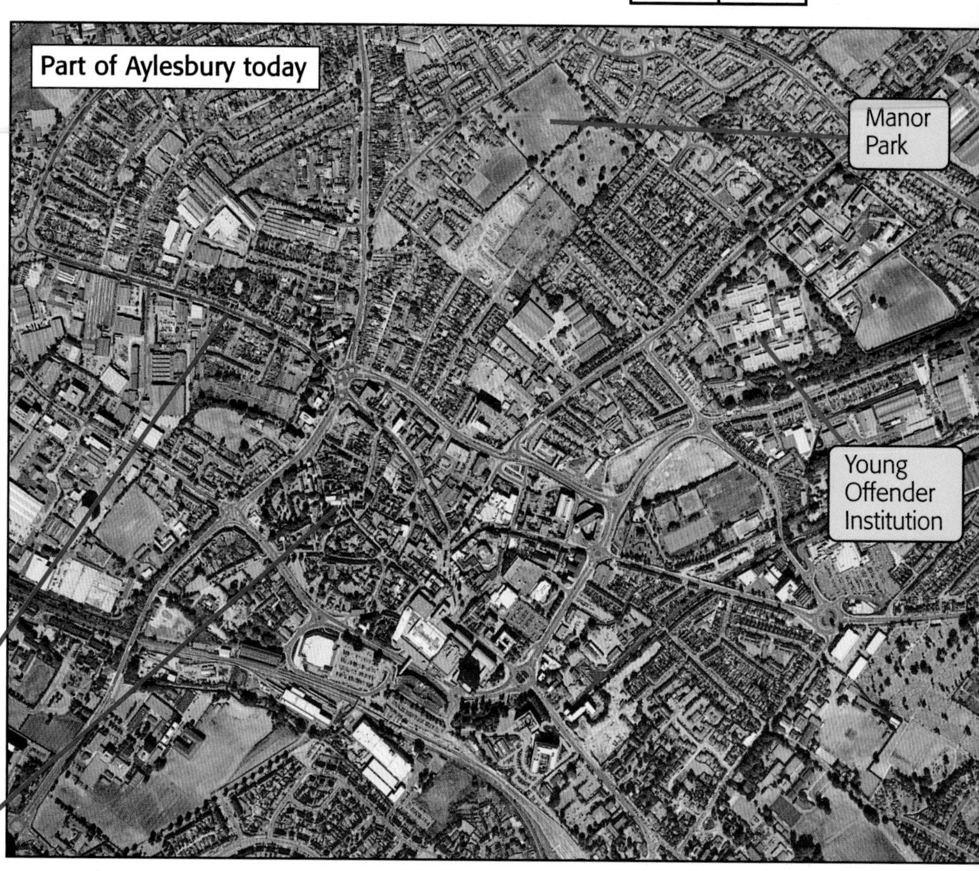

This photo shows just *part* of Aylesbury today, over 1300 years later. It has grown a lot! Turn to the OS map on page 45, and see if you can find the things marked on this photo.

Your turn

1 You are Aelred, a Saxon. Your group has just arrived in the Aylesbury area. You will start a settlement at **X**, **Y**, or **Z** on the map on page 38. Get ready to choose!

a First, make a table like this one.

Factors to consider	Site X	Site Y	Site Z
access to water			
access to timber			
access to farmland			
access to other places			
ease of defence			
safe from flooding			
Total score			

b Now give each site a score of 1 to 5 for each factor. (1 = poor, 5 = excellent.)

c Add up the total scores for each site.

d Which site seems the best choice, to you?

2 In fact Eadwig, your leader, decides on site **X**. So what do you think his main concern is?

3 You've put up your hut and started to farm. Now write a letter to your cousin Oeric back home. Tell him all about your new place. Invite him over!

4 The settlement grew over the years, as the photo shows. Look at your table. Which factors might *not* be so important to the people who live there now? Why?

5 From the OS map on page 45, give grid references for:

a two squares that the Roman road passes through

b the square with Manor Park in

c the square with the Young Offender Institution

6 See if you can identify three more structures (such as buildings or roads or roundabouts) in the photo, and give OS grid references for them.

7 Turn to page 34. It shows Aylesbury too. Look at the big 13th century church, where the Saxon church was.

a Try to give a 6-figure grid reference for the church, on the OS map. (It has a steeple.)

b Can you see anything else in the photo that might have started in Saxon times?

How Aylesbury grew

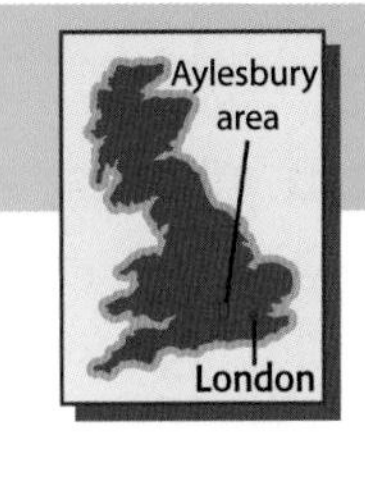

Here you'll see how a settlement can grow, with Aylesbury as our example.

From then to now

The little Saxon settlement at Aylesbury grew slowly at first. Babies were born. New people moved in, to farm, or bake, or make shoes.

The Roman road made it easy to get to. So by 1200 AD, it had become a village with a busy market. People came from all around to buy and sell things.

By 1720, it was famous for breeding white ducks. People walked them to London, to sell them. (For meat and their feathers, which were used in quilts.)

By 1750, Britain was changing. It was the Industrial Revolution! New machines were invented. New factories were set up. Canals and railways were built.

Places like Birmingham and Leeds grew very fast, as people flocked in from the farms to find work in the new factories. These towns grew into cities.

Aylesbury remained a small market town. But even it saw changes: a canal that linked it to London and Birmingham; a railway to Birmingham; a couple of factories.

Then, in 1839, a rail link to London was opened. (So the ducks went by train!) This link to the UK's biggest city helped Aylesbury to grow faster.

In the 1960s it shot up in size. Because thousands of people from London were moved there, to new council housing. (London did not have enough homes.)

Aylesbury still has its market. The old factories are gone, but it has some new industries. It is still changing. And as you'll see later, it's about to grow a lot bigger.

How it has grown since 1830

These two maps show how Aylesbury has grown. Both have the same scale.

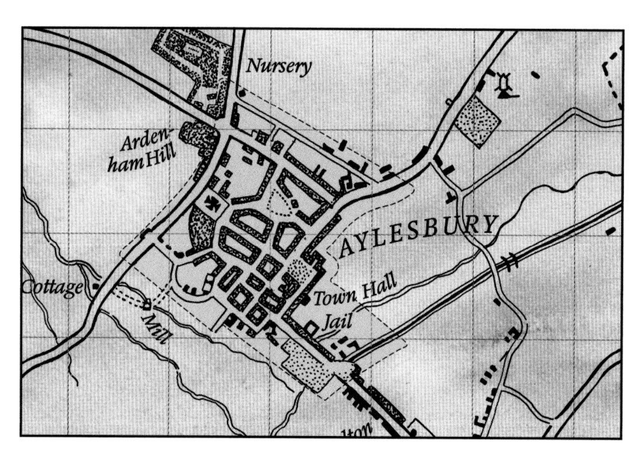

▲ Aylesbury in 1830.

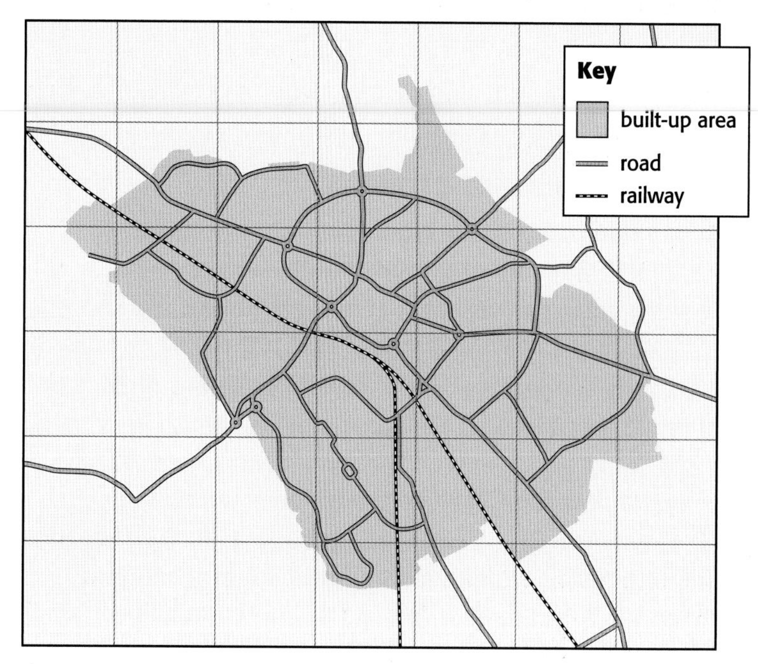

▲ Aylesbury in 2002 (or 172 years later).

Your turn

1 Aylesbury is a *market town*. What does that mean?

2 Look at the map of Aylesbury for 1830. Each square represents 1 square kilometre. You can work out the area *roughly*, by counting squares like this:

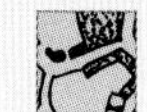

Full = 1.

At least half full = 1.

Less than half full = 0.

About what area did Aylesbury cover in 1830?

3 Now look at the map for 2002. Here each square also represents 1 square kilometre.
  a About what area did Aylesbury cover in 2002?
  b About how many times larger was it in 2002 than in 1830?

4 Look at table **A**.
  a What was the population of Aylesbury in 1830?
  b How many times larger had it grown by 2005?

5 A graph is good way to show population growth.
  a Draw a graph like the one started in **B**, for table **A**. Use a whole page, and complete both axes.
  b If the population keeps growing at this rate, what might it be by 2020? Show this on your graph.
  c Now, on your graph, mark in the events from list **C**.

6 Try to explain how this helped Aylesbury to grow:
  a the opening of the canal
  b condensed milk was made there
  c the opening of the railway to London
  d the opening of Friar's Square shopping centre

C Events that helped Aylesbury to grow

Year	Event
1814	Grand Union canal opens, linking Aylesbury to London and other places
1839	Birmingham Railway opens, linking it to Birmingham
1865	Printers from London set up a printworks here
1870	A factory to make condensed milk starts
1892	Metropolitan Railway links Aylesbury to London
1960	New council houses are built for people from London, where there's a housing shortage
1988	Sony Music sets up factory
1991	New shopping centre opens (called Friar's Square)

A Population of Aylesbury

Year	Population
1810	3400
1830	5000
1850	6000
1870	6900
1890	8900
1910	11000
1930	14400
1950	21200
1970	40500
1990	51000
2005	65200
2010	?

B

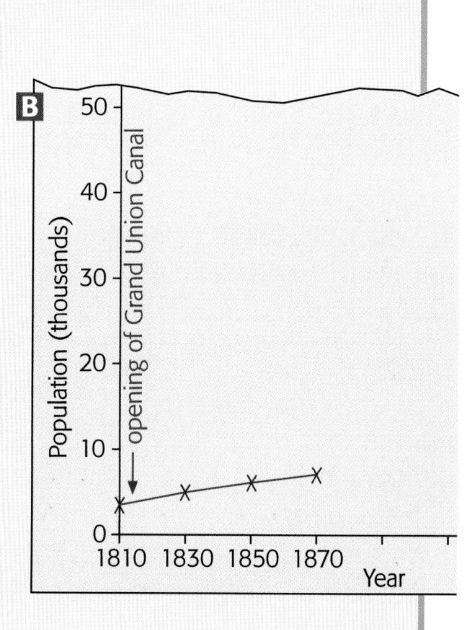

3.4 The pattern of growth

Here you'll learn about the pattern that settlements followed, as they grew. The photos are from Aylesbury.

As a settlement grows ...

- A settlement usually grows out from the centre. So the centre is where the oldest buildings are.
- As it grew, homes in the centre got turned into shops and offices, that people could get to from all directions.
- So the centre became the **central business district** or **CBD**.
- The first factories were built along canals, rivers, or railways, to move goods easily. (The canals, rivers, and railways were near the CBD – so the factories were too.)
- Low-cost terraced houses were often built close to the factories, and canals and railways, for their workers.
- As the population grew, new houses were built at the edges of the settlement, where there was space, and cheaper land.
- Today, new industries usually set up close to main roads, towards the edge of town.

One way to show the pattern

Look at this diagram. It is a **model:** it sums up in a simple way the pattern of growth in our settlements.

No town or city is *exactly* like this model. For example in real life the CBD is never a neat circle!

But many fit it quite well.

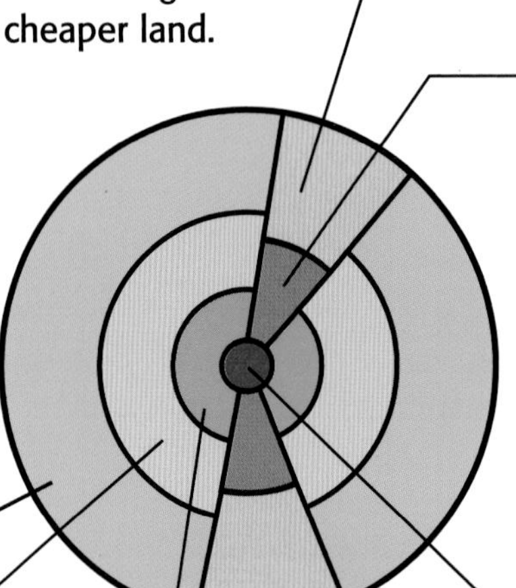

Modern industrial area
It has **Industrial estates** and **business parks**. Mostly built since 1960 – further out, and near main roads.

Old industrial area
From 19th century. Along a canal, railway, or river, near the CBD. Many of the old factories have now gone.

Housing 1960s onwards
Modern housing estates, even further from the CBD. May be near parks and shopping centres. This area is the **outer suburbs**.

Housing 1930s–1950s
Larger houses, with gardens, further from the CBD. May be near parks and shopping parades. This area is the **inner suburbs**.

Old housing near the CBD
Like these old terraced houses. May have corner shops nearby. But lots of old housing has now been replaced by modern flats.

The CBD, in the centre
As well as the main shops, you'll find offices, banks, restaurants, cafes, the main post office, and so on.

In general, as you move out from the CBD:

- land gets cheaper to buy or rent
- housing gets more modern.

Your turn

Look at this map for Aylesbury. It uses the same colours as the model on page 42. But an extra colour has been added, to show an area with really old houses. They date from the 14th to the 18th century. The oldest house in Aylesbury was built around 1386, and is still standing!

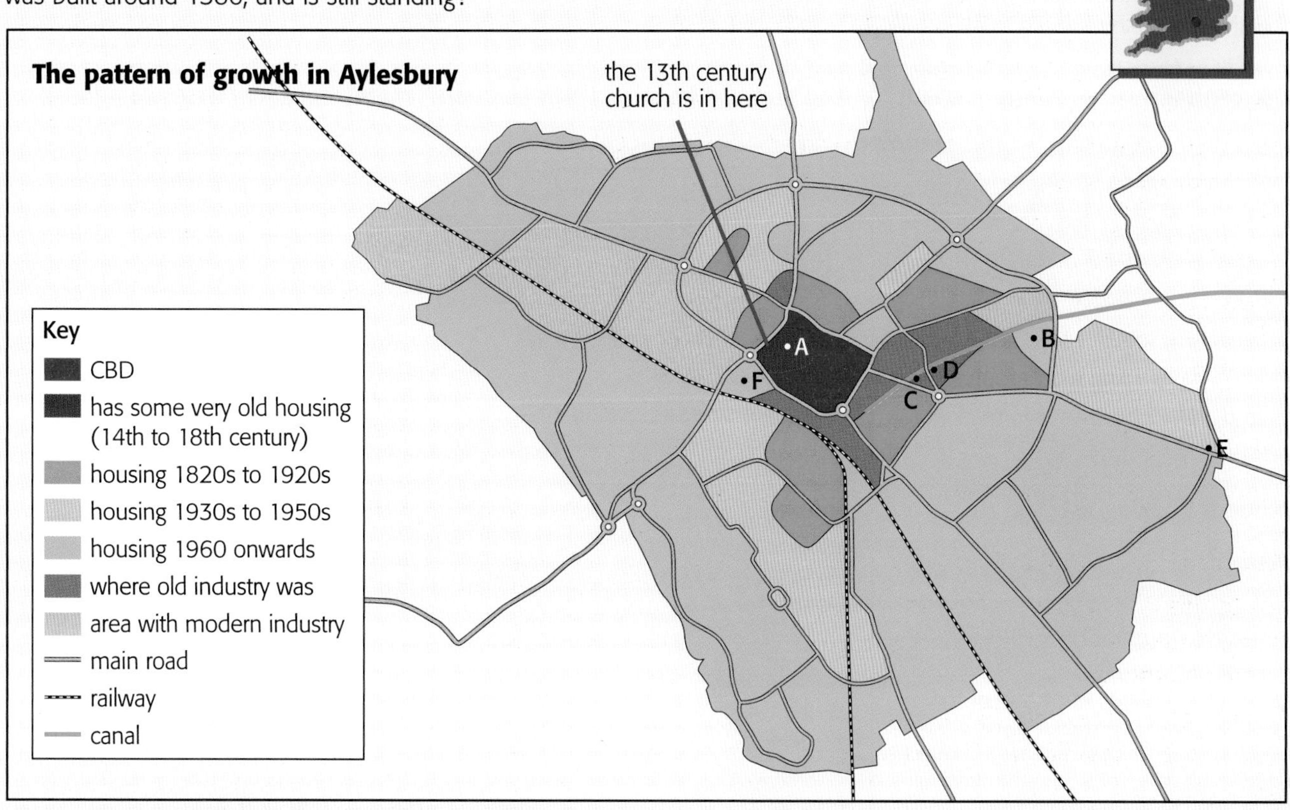

1 The map of Aylesbury above does not look much like our model! But let's see how well it matches it. Say whether each statement is true or false.
   a The CBD is roughly in the centre of Aylesbury.
   b Overall, the oldest housing is towards the centre.
   c The modern housing (1960 onwards) tends to be furthest from the CBD.
   d The old industrial area was close to the CBD.
   e The old industry was along the canal and railway.
   f The modern industry is further out.

2 Look at these houses. One is at **A** on the map above. The other is at **B**. Which house is at which place?

1

2

3 The map shows where the old industry ***was***. But you won't see much sign of it in Aylesbury today.
   a Look at **C**. The condensed milk factory was set up here, in 1870. (See box C on page 41.) The factory is shown in a photo on page 42. Which photo?
   b But the factory was knocked down in 2004. Now there are modern flats at **C**. Do you think it would be a good location to live in? Give reasons.

4 **D** is where the printworks was set up in 1865.
   a Why did the printers choose this site?
   b The printworks has gone. There's a Tesco superstore in its place. Why do you think Tesco chose this site, rather than one at **E**, where land costs less?

5 There is very modern housing at **F** on the map.
   a Does this fit with the model on page 42? Explain.
   b How do you think this housing came to be here?

6 The map shows just the main roads (not all the streets).
   a Many of the roads head towards the CBD.
      i How does this help the CBD?
      ii What problems might it cause?
   b A road loops around the CBD. See if you can explain why this was built.

Be a land-use detective!

The land in settlements is used for many different things: homes, schools, shops, roads…
Here you'll learn how to spot different land uses on an OS map.

Clues from OS maps

OS maps give you lots of clues about what the land in a place is used for.
Like these:

The central business district (CBD)

- the main roads lead to it
- may have a ring road around it
- may have churches, information centre, museums, as well as business

Old terraced housing (19th century)

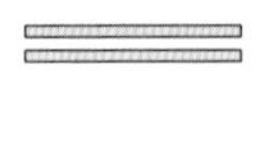
a street of terraced houses with no gardens

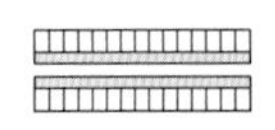
a street of terraced houses with gardens

- straight rows of small houses
- may have gardens, or not
- straight streets close together
- near factory, railway, or canal

Later housing (1930s to 1950s)

wider streets (often curved as here)

the houses are often semi-detached

- usually larger, with gardens
- often in pairs or in short rows
- streets straight or curved
- schools and parks nearby

Modern housing estates

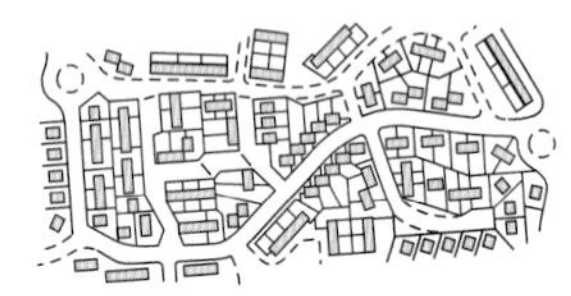

- usually towards the edge of town
- houses usually have gardens
- often in small groups, not rows
- schools and parks nearby

Industrial area

- look for labels **Industrial Estate** or **Ind Est**, or **Works** or **Wks** or **Factory**
- older industry is usually along a canal, river, or railway
- modern industry is usually on or near main roads

Other land uses

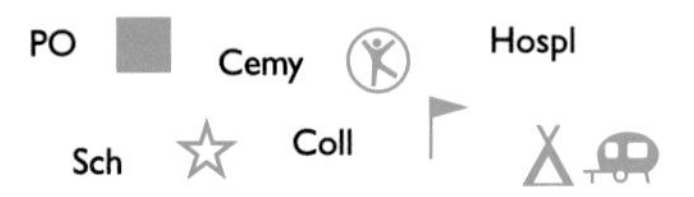

- the map also shows things like schools, post offices, churches, hospitals, farms
- it uses abbreviations or symbols for these, as above (There is a list on page 138.)

Now get ready to be a detective, for land use in Aylesbury.

Your turn

1 Look at the OS map of Aylesbury, on page 45.
  a Identify the CBD, using the clues above to help you.
  b Give four-figure grid references for the squares it is in.

2 Aylesbury is a market town. It's had a market since Saxon days. Which map square is the market in? Use the photo on page 34 to help you locate it. (The top of the photo is north.)

3 Give grid references for two squares with industry.

4 Say as much as you can about the house at 830137.

5 You want to buy a modern house in Aylesbury, with lots of space around it. You drive to London to work every day. There are four houses for sale, at:
A 805133 **B** 823137 **C** 816147 **D** 842136
They are all in good condition.
  a Which one do you think might suit you best? Why?
  b Say why you rejected each of the others.

6 The land within Aylesbury is used for many purposes or **functions**. Look at this table:

Function	Example	Grid reference
housing	Southcourt	
shopping		
education		
health care		
tourism		
leisure / fun		
law and order		
rail transport		

  a Copy the table, and fill in just *one* example for each function. Give a four-figure grid reference.
  b See if there are any other functions you can add to your table, and complete their rows.

Your turn
QUARRENDON CP
Pumping Station (dis)
Weedon Hill
Weedon Hill House
Weedon Hill Farm
Stud
Grendon Hill Farm Cottages
Berryfield House
St Peter's Chapel (remains of)
Medieval Village (site of)
Moat
Earthworks
Civil War Earthworks
Quarrendon House Farm
El Sub Sta
BIERTON WITH BROUGHTON CP (Part of)
Watermead
Hotel
Holman's Bridge
Football Gd
Dunsham Farm
Church Farm
Stone Bridge
Haydon Hill
Quarrendon
Elmhurst
Sewage Works
Factory
ROMAN ROAD
Manor Park
AYLESBURY
Broughton Crossing
Superstores
Industrial Estate
HM Young Offender Institution
Haydon Mill Farm
Grand Union Canal Walk
Victoria Park
TA Centre
Civic Centre
Offices
College
California
Walton
Aylesbury Park Golf Club
Lower Hartwell
Rifle Spinney
Weirs
Southcourt
Barnet's Close
Hartwell House (Hotel)
Park Hill
Park Villa
Walton Court
Bedgrove
Bedgrove Park
Beech Walk
Portway
Sedrup Lane
Peverel Court
Calley Farm
Sedrup Farm
The Firs
Sedrup
The Rectory
STONE WITH BISHOPSTONE AND HARTWELL CP
Midshires Way
North Buckinghamshire Way
Stoke Mandeville Hospital
Amb Sta
Red House Farm
Wendover Road
Ford
STOKE MANDEVILLE CP
Allot Gdns
Beech Cottage
Manor Farm
Hall End
Bishopstone
Fish Pond
Moat
Manor Farm
Standal's Farm
Old Farm
Bishopstone Farm
Brook Farm
Alberta Farm
Stoke Mandeville
Swan's Way
0 0.25 0.5 0.75 1 km
79 80 81 82 83 84
11 12 13 14 15 16

How's Aylesbury doing today?

Every settlement has good and bad points. Here you'll find out about Aylesbury's – and how it plans to improve.

How is it doing?

The settlement at Aylesbury is now over 1400 years old. So how is it doing? Here are some of the things people say about it.

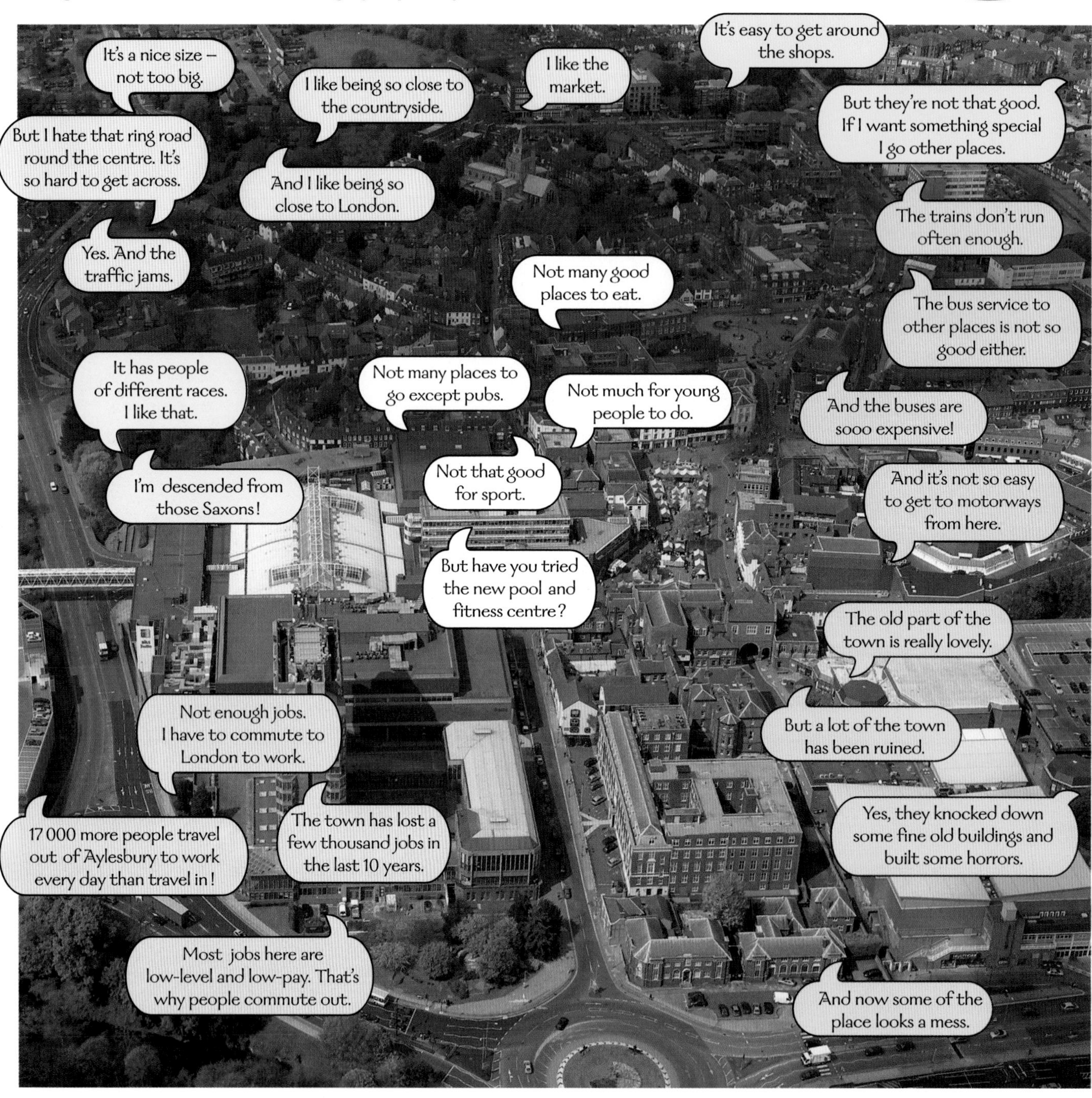

Grumbles

So the people of Aylesbury have some grumbles. They don't think the town is as good as it should be, or could be.

What would *you* do, to improve it?

Improving Aylesbury

Now there are big plans to improve Aylesbury. These are just some of them:

- redevelop the town centre to make it more attractive
- invite upmarket shops to open in Aylesbury
- provide training to improve people's work skills
- get more high-value businesses and industries to set up here
- build a new business park
- improve the road links to the motorways
- improve rail and bus services

It has started already

The work on Aylesbury has started already. Look at this example:

This shows one end of Exchange Street in 2005, with the canal in the background. Not too attractive.

The area is being **redeveloped**. This shows part of the redevelopment: a new theatre.

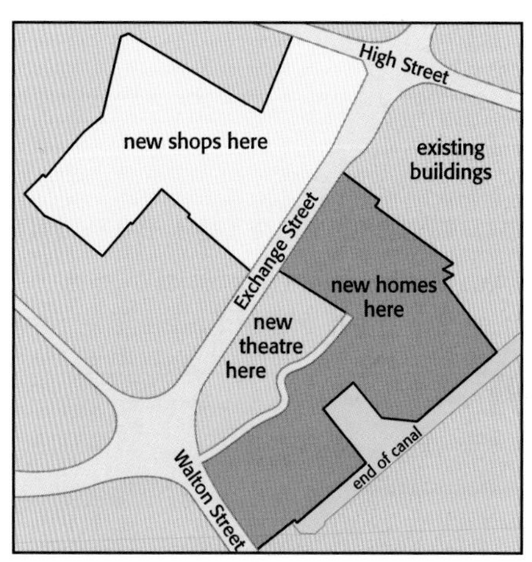

The redevelopment also includes homes by the canal, and shops, and restaurants. Look at the plan.

What do you think? Is it an improvement?

Your turn

1 Here are some terms used in this unit:
  a commute b high-value businesses
  c business park c redeveloped
  Write down what they mean. (Try the glossary?)

2 People like lots of things about Aylesbury – but as you see they also have some grumbles.
  Pick out one grumble about Aylesbury as a centre for:
  a transport b tourism c shopping

3 People say Aylesbury may become a **dormitory town**: people sleep there but commute out every day, to work in other places.
  a What is the evidence that this may be happening?
  b What is causing it?
  c Do you think it is good, or bad, for Aylesbury? Give as many reasons as you can.

4 The council is doing some things to stop Aylesbury becoming a dormitory town.
  a See if you can pick these out from the list at the top of the page.
  b But some things in the list will make it *easier* for people to commute to other places. Which things?

5 Look at the Exchange Street redevelopment above. (It's around 822136 on the OS map on page 45. Exchange Street links the A41 and A413.) It is a response to some of the grumbles. Which ones?

6 So what about *your* settlement, or local area?
  a Give as many good points about it as you can.
  b Now give some grumbles.
  c Choose one grumble and say what *you* would do to solve the problem, if you were in charge.

A new challenge for Aylesbury

The UK needs a lot more new homes, and fast. In this unit you'll find out why – and explore how Aylesbury will be affected.

Help, we need more homes!

The UK is short of homes. There just aren't enough to go round. So house prices have gone sky high, as people compete to buy them.

And that means more and more people can't afford a home. They end up renting a flat, or living with mum and dad, when they really want to buy a place of their own.

Now the government says we must have 3 million more homes by 2020. In other words, 240 000 new homes a year, on average, from now until 2020. Will one of them be for you?

From news reports, 2007.

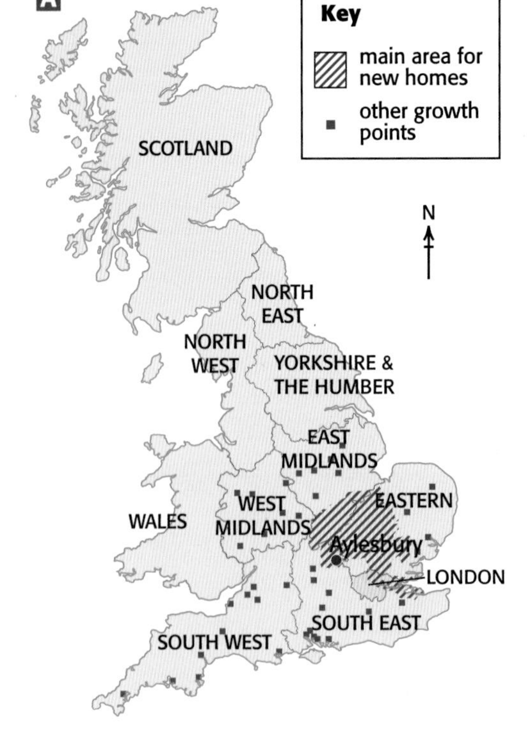

So where will the new homes go?

Map **A** above shows where most of the new homes will go. Aylesbury is in the main growth area. The new homes will be on two kinds of sites:

Some will be on **brownfield sites** – sites in towns and cities that were built on before. Like on the site of this old flour mill, beside the canal in Aylesbury.

2

The rest will be on **greenfield sites**. These are sites that have not been built on before. Like on this farmland on the edge of Aylesbury.

Aylesbury's new homes

Aylesbury is to have lots of new homes: around 15 000 more by 2020. And that means about 30 000 more people! It's a big challenge.

Map **B** shows the first sites chosen for these homes. It shows a proposed new business park too.

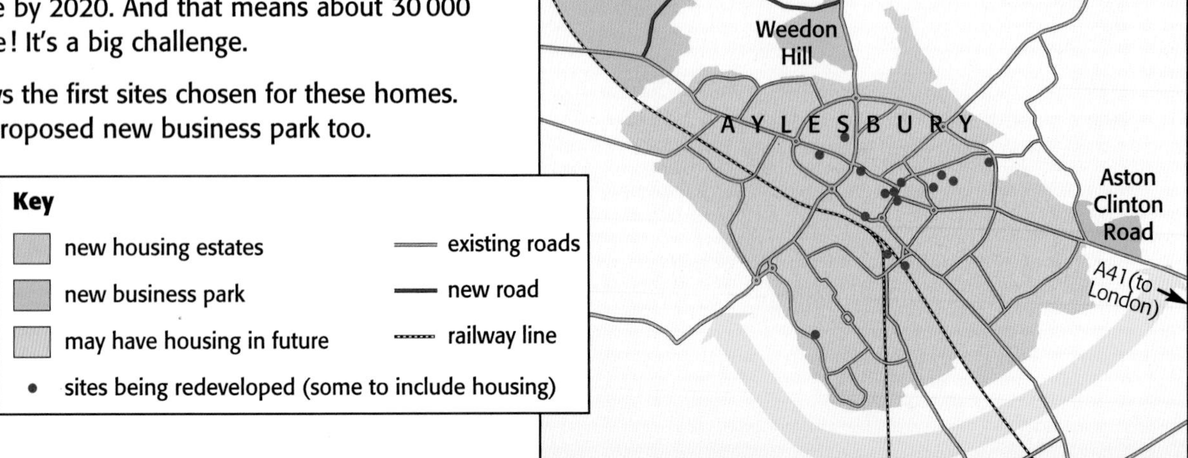

Your turn

1 Why does the UK need lots more housing? Each drawing above shows one reason.
  a Study each drawing, then put the reason into words.
  b Now see if you can think of any other reasons.

2 Map **A** on page 48 shows where new homes will go.
  a Where in Britain is the main area for them?
  b See if you can explain why. Pages 70 and 74 may help.

3 Now suggest reasons why Aylesbury has been chosen for new homes. The maps on pages 45 and 48 may help.

4 Explain what these terms mean, in your own words:
  a brownfield site
  b greenfield site

5 There is conflict across the UK, about using greenfield sites. Look at these people:

  a In this group, who is **A** likely to be in conflict with?
  b Who might side with **A**, but for a different reason?
  c Which two might get on really well together?

6 Now compare the two photos on page 48. Which type of site would *you* choose first? Explain why.

7 Map **B** on page 48 shows some sites for Aylesbury's new homes. Compare it with the OS map on page 45.
  a Is this a greenfield site or a brownfield site?
    i Berryfields   ii Weedon Hill
    Hint: look for those names on the OS map.
  b Where will *most* of Aylesbury's new homes be?
    i on greenfield sites ii on brownfield sites
    (Check the key on map **B**.)
  c The Berryfields and Weedon Hill sites are quite close. Why don't they join them to give one big site? The OS map has clues!

8

Aylesbury will have over 30 000 more people by 2020. It will need more doctors, to look after them. List other people will it need more of, to look after them. (Teachers? Police?) See how many you can give.

9 When the Saxons arrived in the Aylesbury area in 571, they were looking for a place with water, and wood for fuel, and easy access to other places. You are a developer. You are building new homes like this one in Aylesbury, for all the people who'll arrive soon:

  a What kind of fuel will you provide for them? Where do you expect to get it from?
  b How will you provide water? Where from?
  c What will you do, to help with access to other places?
  d Map **B** shows that one of those needs (for fuel, water, access) has been thought about already. Which one?

10 You were a Roman commander, in charge of building roads in Britain. Tell us your opinion about the site of the proposed new business park in Aylesbury (map **B**).

Sustainable development for Aylesbury

In this unit you'll learn what sustainable development means. We'll use the new housing at Aylesbury as an example.

What is sustainable development?

Sustainable development means development that will improve our lives, and not lead to problems in the future. It has three strands:

THE THREE STRANDS OF SUSTAINABLE DEVELOPMENT

Can the town afford this development?

And will there be jobs around here for us?

1 ECONOMIC
to do with:
- money and wealth
- earning a living

Will it make life easier for us ...

... and help us to be healthy and happy?

2 SOCIAL
to do with things like:
- how we live our lives
- what we do in our leisure time
- our health
- our education
- our relationships with others

Will it help the natural environment ...

... or at least not harm it?

Will the buildings look good?

3 ENVIRONMENTAL
to do with:
- our impact on the natural environment (air, rivers, climate, wildlife, ...)
- the quality and look of our built environment (homes, streets, railway stations ...)

What if ...
- ... nearly all the countryside got built over?

Look at all the questions. We need to ask questions like these about any planned development – such as a new housing estate.

If all the answers are *yes*, the development is likely to be sustainable. It can be carried on safely, without causing problems in future. But if any answer is *no*, watch out. There are problems ahead!

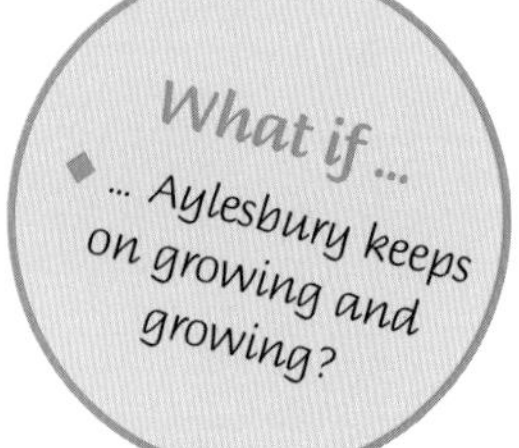

The new developments at Aylesbury

The Aylesbury planners want the new housing developments to be sustainable. Look at their plans.

1 ECONOMIC

- The District Council will buy the land from the farmers and others who own it. The government will give money to help us do this.
- We'll ask **developers** to build the new homes.
- Then they will be sold or rented out.
- We will make sure that some homes are lower cost, for young people and people on lower wages.
- We are also working hard to bring new businesses to Aylesbury, to provide more local jobs for people.

2 SOCIAL

- There will be homes to suit people with different needs – single people, elderly people, families with children.
- We'll make sure there are enough school places, and doctors, and so on.
- There will be a good bus service into town.
- And cycle paths everywhere.
- In town, there will be better shops, more cafes and restaurants, a new theatre, and lots more new facilities. (There's a new swimming and fitness centre already.)

3 ENVIRONMENTAL

- The new homes will look good, and be good quality.
- There will be trees and open spaces around them.
- They will waste very little energy. For example they will be well insulated, to save on heating. (Saving energy helps us to fight global warming.)
- The good bus service, and cycle paths, will encourage people to use their cars less.

▲ *One of the new housing estates at Weedon Hill will look like this.*

Your turn

1

School to be rubbish!

Your town has run out of places to put rubbish. The council plans to knock down your school, and use the site as a **landfill site** (rubbish dump). This is a development – but is it sustainable? Decide, and give your reasons.

2 Look at the list on the right. It's about a great new housing estate. This is your task:

i Make a larger copy of the Venn diagram below.

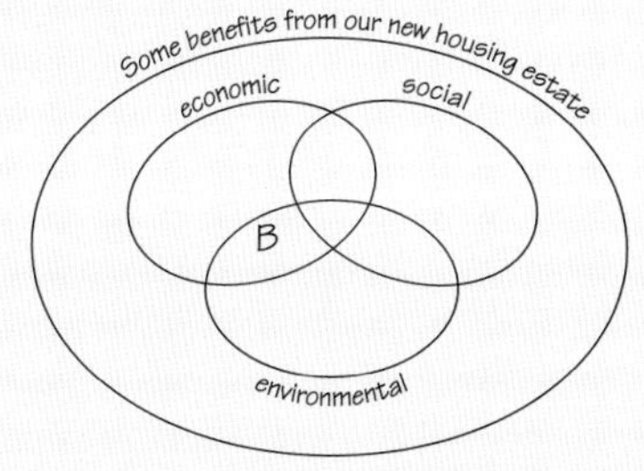

ii Think about each statement in the list. Then write its letter (**A** to **L**) where you think it should go, in your Venn diagram. (One has been done for you.) If you think it belongs to two loops (or even three) write it where they overlap.

Our new housing estate

- **A** They have bins for recycling everything.
- **B** Our electricity bills are really low. That's because the houses are designed to save energy.
- **C** There are lots of birds in the trees around the house.
- **D** The paths and lawns are looked after so well!
- **E** There's a good little supermarket on the estate.
- **F** There's a club for young people, with lots of activities.
- **G** My dad got a well-paid job in the new business park, just a couple of miles away.
- **H** My room is nice and big so friends can come and stay.
- **I** It's far cheaper living here than in our old place in London.
- **J** They've put on a good cheap bus service into town.
- **K** I've made loads of good friends here already.
- **L** I can walk to school. I really like that.

3 Now imagine another new housing estate. This one is an example of *unsustainable* development.

a Make up ten statements about it. One could be: *Miles from everywhere, and no buses.*

b Number them 1 to 10.

c Swop statements with your partner.

d Then repeat question **2** for these new statements. But this time your Venn diagram should have the label *Unsustainable development*.

4 Let's go shopping!

The big picture

This chapter is about shopping, and how it is changing. These are the big ideas behind the chapter:

- Shopping is all tied up with geography!
- We're willing to travel further for some goods than others.
- Shops need to set up where they'll get enough customers, in order to make a profit.
- Shopping is always changing. Out-of-town shopping and internet shopping are examples of changes.

Did you know?
- Until the 17th century, most 'shops' were market stalls.

Did you know?
- 200 years ago, all your clothes would have been hand sewn ...
- ... because sewing machines were not invented until around 1850.

Your goals for this chapter

By the end of this chapter you should be able to answer these questions:

- What are convenience goods?
- What are comparison goods?
- For which of those goods are people willing to travel further?
- Why do shops set up in some places – and not others?
- Why do bigger settlements have a larger range of shops than small settlements do?
- What is an out-of-town shopping centre, and what am I likely to find there?
- Who may benefit from out-of-town shopping – and who may lose out?
- How does internet shopping work?
- Who may benefit from internet shopping – and who may lose out?

What if ...
- ... internet shopping took over?

What if ...
- ... chain stores were banned?

And then ...

When you finish the chapter, come back to this page and see if you have met your goals!

Your chapter starter

Look at the first photo on page 52. That's the shop owner at the door, with the tape measure around his neck.

About how many years ago do you think the photo was taken:
30? 60? 100? 120? 180?

What do you think shopping would have been like, in that shop?

In what ways would it have been different from in the other shop?

4.1 Shopping around

In this unit you'll learn why shops are where they are – and see how shopping is all tied up with geography!

Did you know?

In the UK we spend:
- about £8 billion a year on food shopping
- about £3 billion a year on clothes and shoes.

Just good fun?

Shopping can be fun. But behind the fun is some very serious business! Shops don't pop up just anywhere. There is a pattern to where they are located. And this pattern is the result of the two key factors below.

1 There are two types of goods

There are some things we buy often, or quite often, that don't cost much, and which we are happy to buy…

… in the nearest convenient place, for example in the local corner shop. So these good are called **convenience goods**.

Supermarkets sell convenience goods, so it's handy to live near one. You can also buy them in the shops in petrol stations.

Then there are goods that cost more, and we don't buy so often. We like to compare styles and prices before we buy …

… so they are called **comparison goods**. We are prepared to travel to get a good choice – for example into the town or city centre.

If we want something really special, like a dress for a very special event, we might even go to a city some distance away.

2 Shops have to make a profit!

Profit is the money left after you pay all the costs of your business.

If you sell convenience goods, you could make enough profit even in a village. Local people may call in several times a week.

But if you try to sell the latest fashions there, it's a different story. You won't get enough customers. Your business will fail.

You will need to move your clothing store to a place where there are lots of shoppers. For example into the next town.

So let's see how all this affects you, when you go shopping.

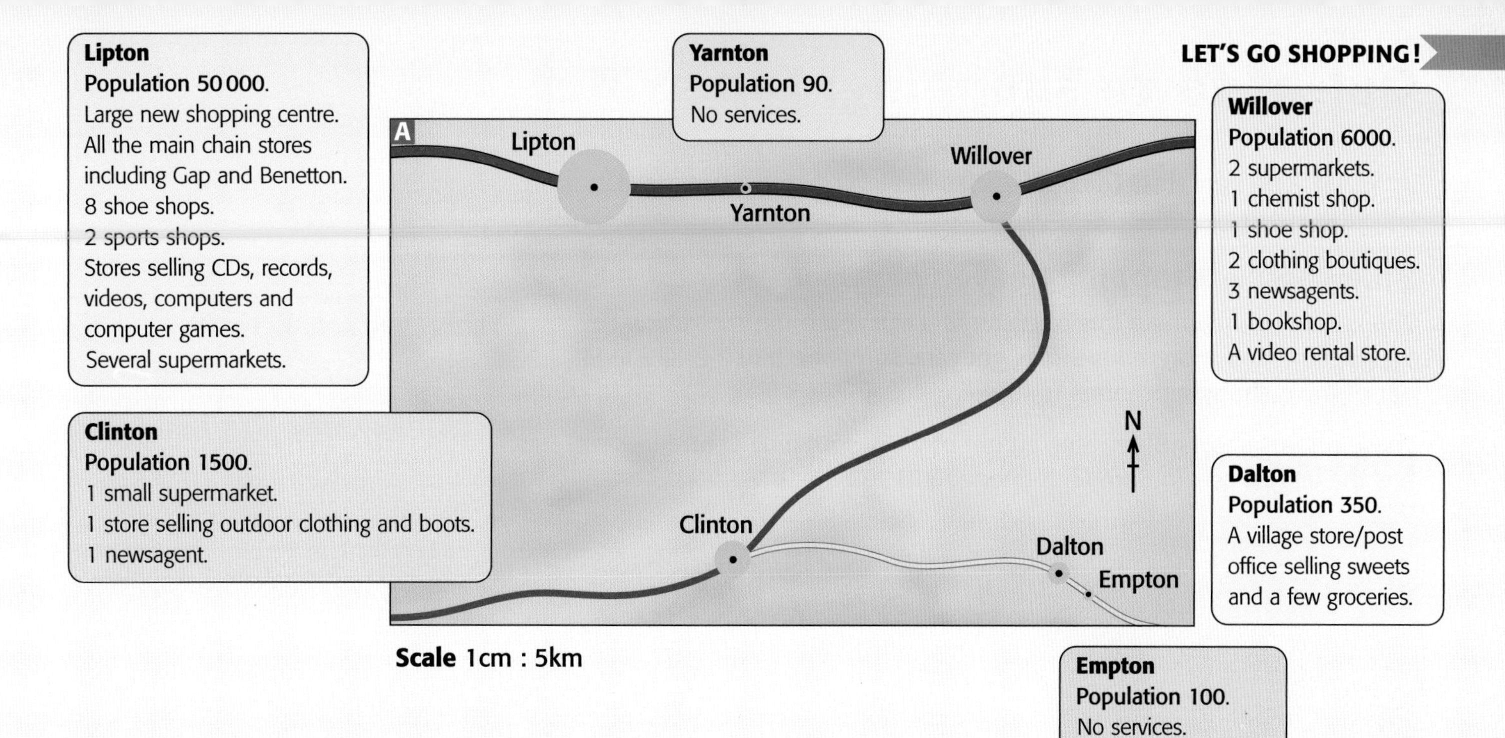

Scale 1 cm : 5 km

B

Item	Cost	Frequency – how many times a year I buy it	Type of good	Where I will buy it	Distance travelled
T-shirt					
Crisps					
Hair gel					

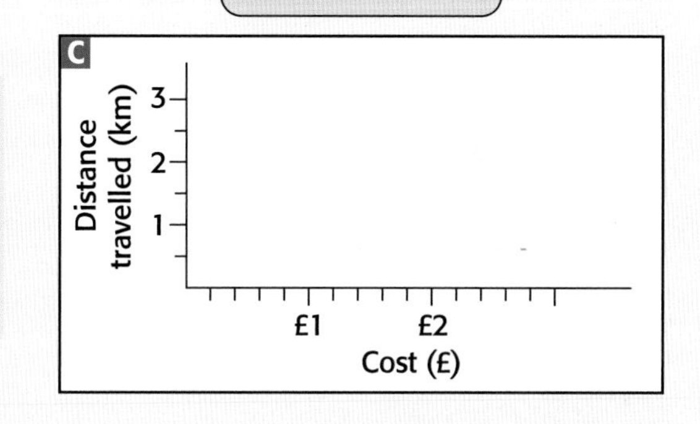

Your turn

Look at map **A** above. You live at Empton on it – and you are going shopping!

1 Look at table **B** started above. It shows items you might buy *at least once a year*.
  a Make your own table with the same headings, and ten rows to fill in.
  b In column **1**, fill in ten items. Try for a mix of low-cost items and more expensive ones.
  c Write the cost of each item in column **2**. If you are not sure, ask a friend, or guess.
  d Now fill in column **3**.

2 a Explain in your own words what these terms mean:
  i convenience goods ii comparison goods
  b Now, for each item in your table, write the correct term, *convenience* or *comparison*, in column **4**.

3 Next, decide where to shop. You will make a separate trip for each item. (Not like real life!)
  a Start with the first item in your list. Where on the map will you buy it? (You may want a good choice, but at the same time you do not want to spend too much time or money travelling.) Write your decision in column **5**.
  b How far will you travel (*both ways*) for it? Measure along the road, from your Empton dot to the centre of the place, using a suitable method and the map scale. Write your answer in column **6**.
  c Repeat **a** and **b** for the other items in your table.

4 Now you will draw a scattergram. You will plot the cost of items against the distance you're prepared to travel.
  a Draw the axes as shown in diagram **C**.
  b Mark the crosses for comparison goods in one colour, and convenience goods in another.
  c Describe any pattern you find, and then explain it.

5 a Look at map **A** again. Why is there:
  i no supermarket at Dalton?
  ii no video rental shop in Clinton?
  b Why have so many shops set up in Lipton?
  c Now write a general rule linking the size of a place with the number and range of shops in it. (It could start like this: *The larger a place is, the …*)

6 The **sphere of influence** of a place means the area around it, that is affected by it. For example the area from which it attracts shoppers.
  a Which places on the map above do you think are in Lipton's sphere of influence?
  b Which are in Dalton's sphere of influence?

4.2 Out-of-town shopping: Bluewater

In this unit you'll learn about Europe's biggest out-of-town shopping centre, and explore its impact.

Bluewater – shopping heaven?

Shops want to be where lots of customers can reach them easily. So, as more and more households got cars, someone had a bright idea: out-of-town shopping centres!

These photos shows Bluewater, an out-of-town shopping centre in Kent. It is the largest one in Europe. It opened in 1999.

Things you can do there

- shop!
- eat all kinds of food
- go to the cinema
- go boating and cycling
- listen to live music
- join workshops (for crafts, for example)
- sit and think, in a special quiet room
- talk over problems with a chaplain

Outside

- six man-made lakes
- parks to walk in
- a water garden
- parking for 13 000 cars

Bluewater factfile

- It cost £350 million to build.
- It employs around 7000 people.
- At least one busload of shoppers arrives every minute.

Inside

- over 330 shops
- over 40 places to eat and drink
- a 13-screen cinema
- a crêche for children aged 2–8

Before

- Bluewater is built on the site of an old chalk quarry.

Bluewater on the map

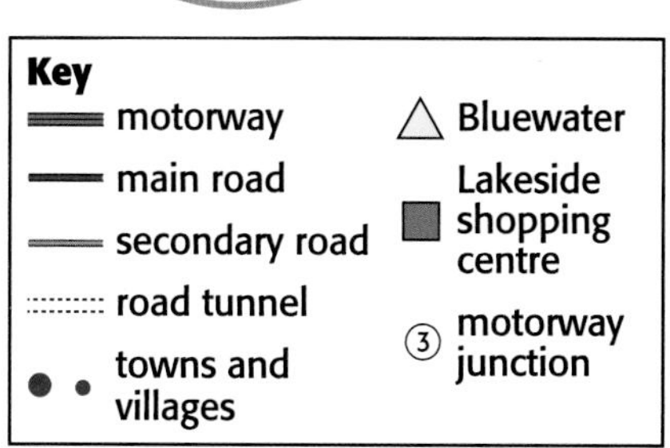

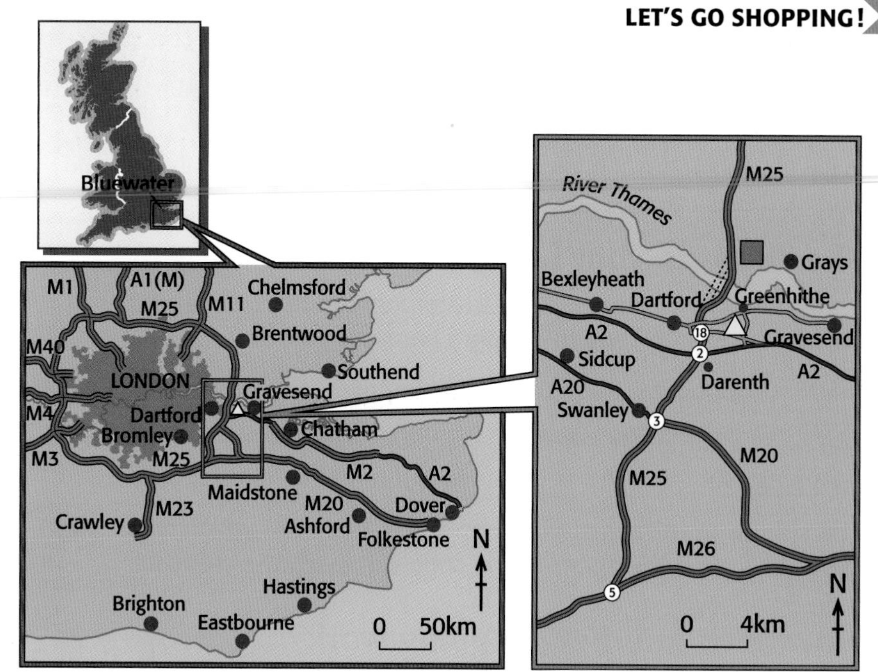

Are there more?

The UK has 11 mega-centres like Bluewater.

They are owned by **developers** – companies who buy land and put up buildings, and then rent them out.

Now the government is worried about their effect on nearby towns, and on traffic. It has not given permission to build more.

The boating lake at Bluewater.

Your turn

1 Imagine you spent the day at Bluewater. Did you enjoy it? Write an e-mail to a friend about your day.

2 a Bluewater was built by *developers*. What are developers?
  b Look at the photos and maps for Bluewater. Why did the developers choose this site? Think of at least three reasons.
  c Why didn't they build Bluewater in London?

3 Every new development affects an area. Some people gain, some lose. Look at the maps. How do you think Bluewater will have affected:
  a a small dress shop in Gravesend?
  b shops in central London?
  c the Lakeside shopping centre?
  d a coffee shop at Greenhithe railway station?
  e a small newsagent's in Darenth?
  f traffic on the A2 and M25?
  Give reasons for your answers.

4 You work for the developers. Make up a leaflet to give local people, to say how Bluewater is helping the area. (Don't forget jobs, and what the site was like before.) Give your leaflet a snappy title.

5 You live in a town near Bluewater: Bexleyheath. The shops there have lost customers to Bluewater. They want you to save them!
  a Think up ways to attract shoppers back to the town.
  b Then prepare a speech to make to the town's Chamber of Commerce, giving your ideas.

4.3 Shopping on the internet

In this unit you'll learn how internet shopping works – and explore the pros and cons for different groups of people.

The shops in your home

Shops need to be where lots of customers can reach them easily. So why not set up shop in people's homes!

Internet shopping is a big change in shopping. And it is catching on fast.

You need a computer – just like you need transport for out-of-town shopping. But soon you might need just a mobile phone.

The **sphere of influence** of a shop is the area from which it can draw customers. With the internet, one shop can reach the world!

▲ *And we'll have three of those, thank you!*

How does internet shopping work?

The internet is a network of millions of computers around the world, all linked together. You connect your computer to the network using a phone line.

Suppose you want to buy a book from a company called Bookworm, in New York. This is how it works:

Key

- computers in homes, schools and offices
- large computers run by companies called **service providers**. They pass messages around the internet.
- the route your order could take
- links between computers

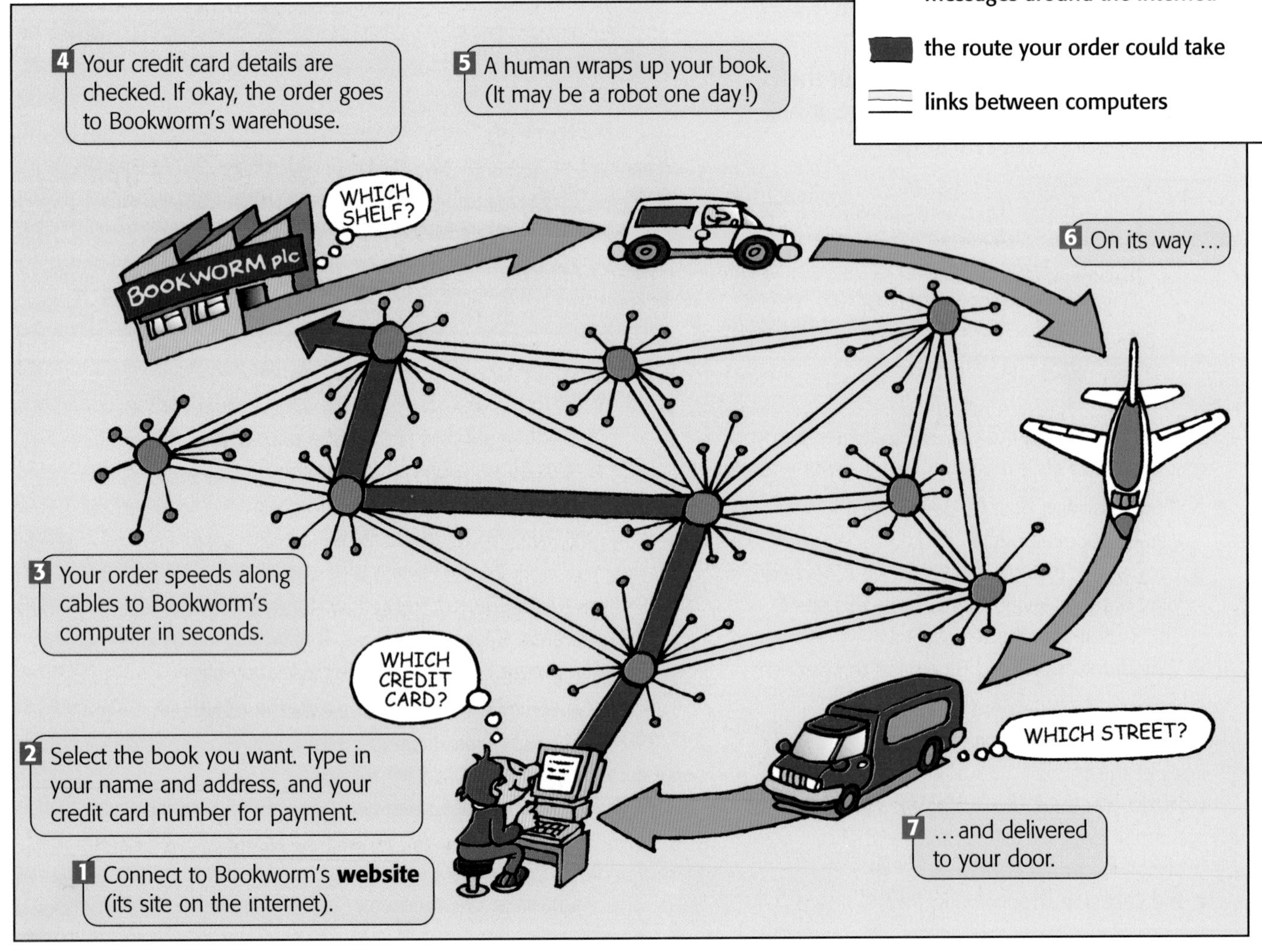

What else can you buy?

You can buy almost anything over the internet, from anywhere in the world. Including food from the supermarket down the road.

▲ *The warehouse behind that website (German branch).*

▲ *Groceries delivered to your door.*

Your turn

1 Explain what the *internet* is, in your own words.

2 What would you say is the *main* difference between internet shopping and ordinary shopping, for:
- a the shopper?
- b the company selling the goods?

3 Look at each item in the list below. Would you be happy to buy it over the internet? Give reasons.
- a a computer game
- b designer jeans
- c a week's supply of groceries
- d a new house
- e a packet of crisps

4 You want to set up a new travel agency. Your choice is:

A **B**

A: rent a shop in the middle of the city, and do it up nicely

B: create a website and sell tickets and tours over the internet

- a Which do you think would cost less to run? Why?
- b Which would give you a larger sphere of influence? How big could this sphere of influence be?
- c Which will let you sell tickets more cheaply? Why?

5 Every day, more and more people go internet shopping.

How do you think this will affect:
- a the local corner shop?
- b a travel agent's in town?
- c an out-of-town shopping centre like Bluewater?
- d air pollution?
- e the number of jobs in shops?
- f companies that deliver parcels?

6 Explain why internet shopping could make life easier for:
- a someone living in a rural area
- b a disabled person
- c a mother with young children
- d a person who works very long hours.

7 Imagine a future where all shopping is on the internet.
- a What effect do you think this would have on:
  i our town centres? ii our enjoyment of life?
- b Do you think this change would be *sustainable*? (Glossary?) Give your reasons.

5 Exploring Britain

You have just arrived in Britain, from the other side of the world. What kinds of things will you find here?

▲ *Some famous landmarks …*

▲ *… a Royal Family …*

▲ *… big cities …*

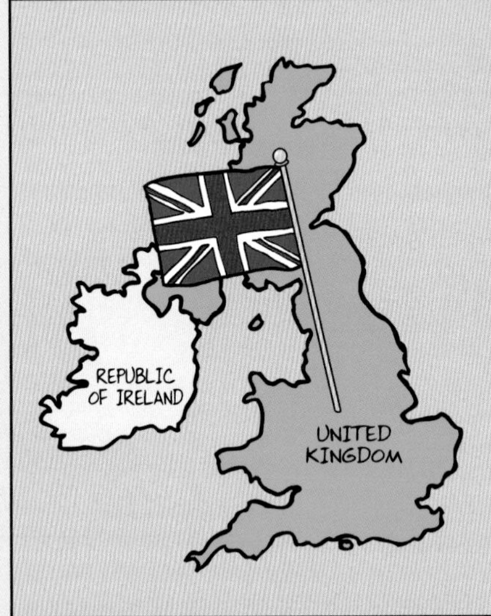

▲ *… small towns and villages …*

▲ *… an exciting coastline …*

▲ *… mmm …*

▲ *… friends, and good times, you hope.*

The big picture

This chapter is about Britain – where you live. These are the big ideas behind the chapter:

- About 10 000 years ago, nobody lived here, because the land was covered in thick ice.
- But as the ice melted, people began to arrive. Over the centuries more and more came, from different places. People are still arriving.
- We have spread all over Britain.
- We have carved it up like a jigsaw, into different regions.
- It is a place of contrasts. Some parts are colder and wetter than others. Some are more crowded. Some are more wealthy.
- And it is still changing.

Your goals for this chapter

By the end of this chapter you should be able to answer these questions:

- Which countries and nations make up the British Isles?
- What are the main physical features of Britain?
- What's the weather like in Britain?
- Who are the people of Britain descended from?
- Which parts of Britain are the most crowded? And least crowded?
- What do these terms mean?

  urban area *rural area* *population density*

- Which are the UK's biggest cities, and where are they? (Give at least six.)
- What kinds of work do people in the UK do?
- What do these terms mean?

  economic activity *primary sector* *secondary sector*
  tertiary sector *quaternary sector*

- Which parts of Britain are wealthiest, and which are least wealthy?

And then ...

When you finish the chapter, come back to this page and see if you have met your goals!

Did you know?

- Britain is the eighth largest island in the world.

Did you know?

- The land that's now Britain once lay at the equator.

Did you know?

- Britain was once joined to France.
- But it got cut off by rising seas, about 8000 years ago.

Your chapter starter

Look at the photos of Britain, on page 60. What's Britain?

Do you think these photos give a good idea of Britain?

Would you leave any out?

What would you add in?

What image comes to *your* mind first, when you think of Britain?

5.1 Your island home

In this unit you'll learn about Britain's main physical features.

The British Isles

They were shaped by hot currents inside the Earth. And by ice, rivers, wind and waves. Just look at them now!

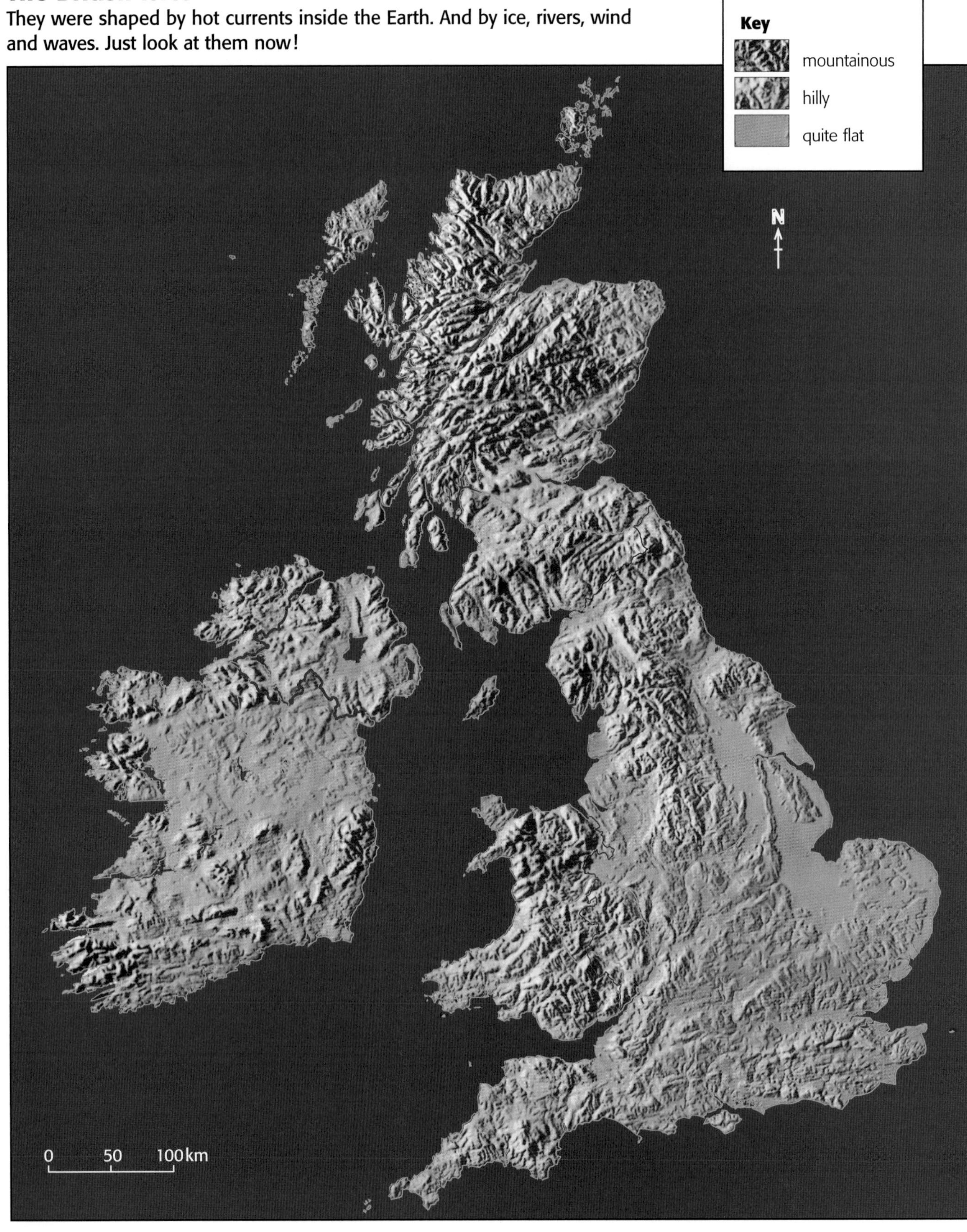

▲ *You'll find places like this in Britain ...*

▲ *... and places like this.*

Your turn

1 Look at the islands on page 62.
  a Point to the one you live on. What is its name?
  b See if you can show exactly where you live on it.
  c Where are the highest mountains, on your island?
  d Where is the flattest land on it?

2 Now look at this map.
  a What do you think the dark yellow shows?

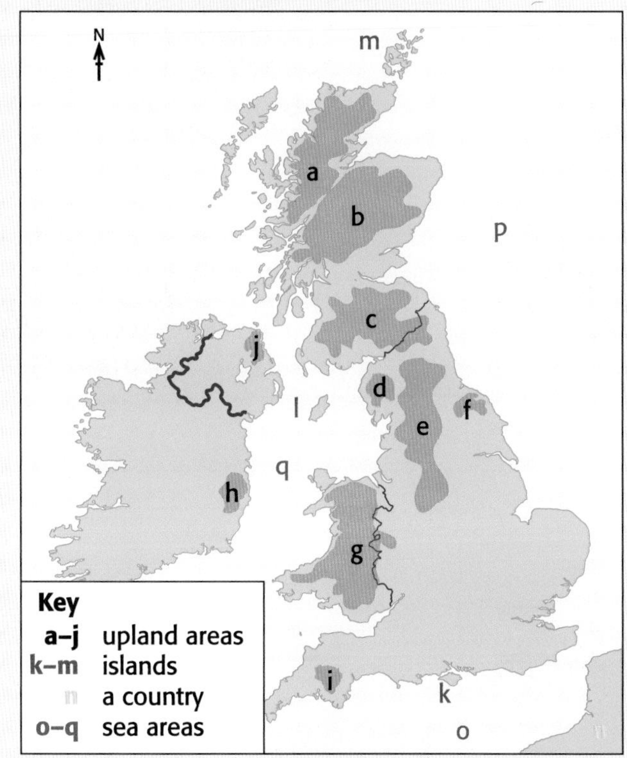

  b See if you can name all the places and features marked on this map. Page 139 will help.
  Start your answer like this: *a* = ____________

3 There are thousands of rivers in the British Isles. Turn to page 139, and see if you can identify the rivers in this list.
  A It's the longest river in Britain. It rises in Wales.
  B This one flows by the Houses of Parliament.
  C Stoke-on-Trent sits on this river.
  D Newcastle sits on this one.
  E This one runs along part of the border between England and Scotland.
  F Did Aberdeen get part of its name from this?
  G This one flows to the Wash, on the North Sea.

4 Photos **X** and **Y** above were taken at **A** and **B** on this map.

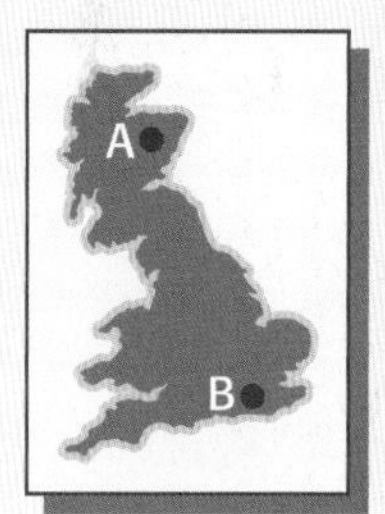

  a Which photo was taken at which place? Explain your choice.
  b Both places were shaped by nature.
    i Which one also shows signs of being shaped by humans?
    ii What do you think it may have looked like, before humans arrived?
  c Write a paragraph comparing the two places. Say what's similar about them, and what's different.

5 You live on an island. Is that a good thing?
  a Make a list of advantages of living on an island.
  b Now list any disadvantages you can think of.
  c Which win, the advantages or disadvantages?

6 Finally, write a paragraph saying where on the Earth the British Isles is. Pages 140 – 141 will help. Include these terms in your paragraph:
equator *ocean* *continent* *Arctic Circle*

It's a jigsaw!

In this unit you'll see how we humans have carved up the British Isles.

Building borders

8000 years ago there were no borders in these islands – because hardly anyone lived here.

But over time, different tribes arrived. They fought over things like land, trade and religion.

In the end, borders were built between different areas. We still have them today.

Two countries

Today, the British Isles is divided into two **countries**: the United Kingdom (UK) and the Republic of Ireland. The UK is the green part on map **A** below.

The UK is in turn made up of four **nations**: England, Scotland, Wales and Northern Ireland.

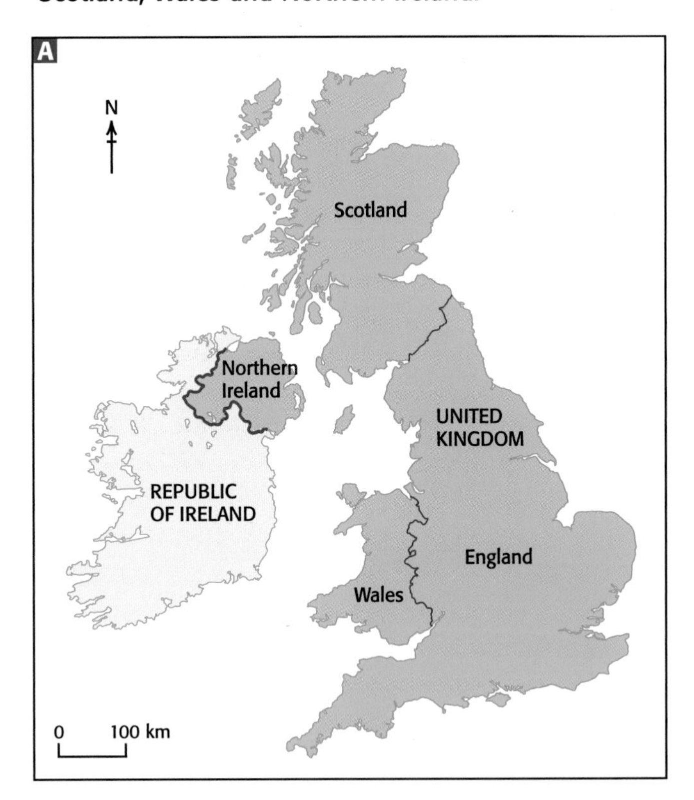

But that's just the start of the jigsaw. For example England is divided into the **regions** on map **B**.

These are in turn divided into smaller areas. Map **C** gives an example. Each area looks after its own services, such as schools, hospitals, and the police.

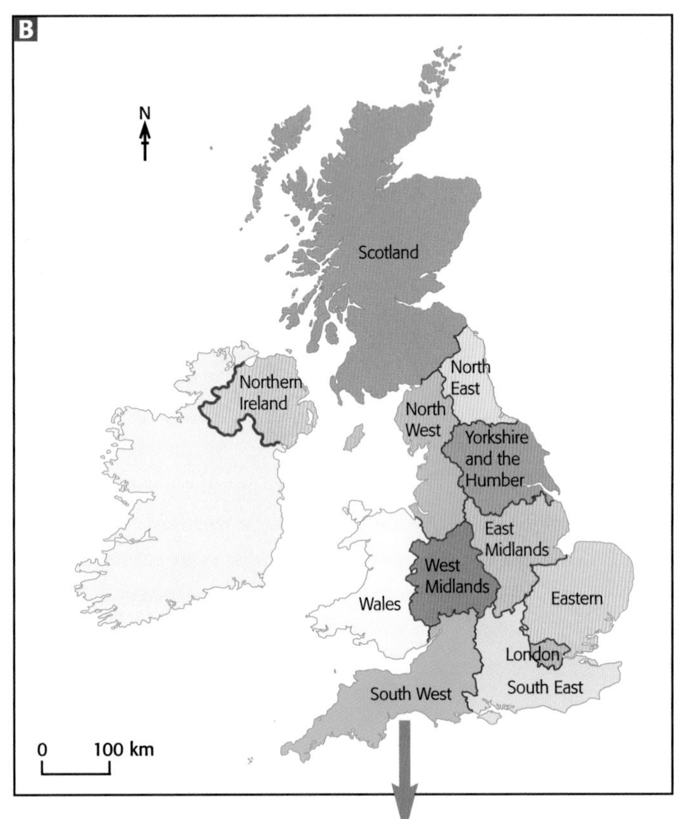

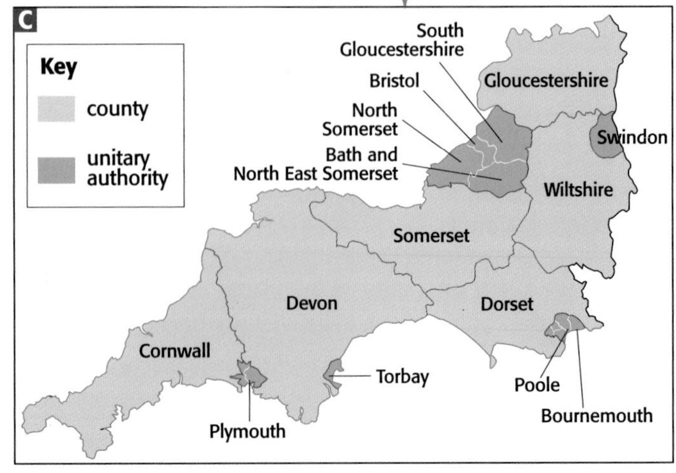

Remember!

the British Isles

the United Kingdom

Great Britain (or just Britain)

Some facts about the British Isles

UK / Republic of Ireland					
Area (square kilometres)	130 400	77 100	20 800	14 200	70 300
Population (millions)	50.5	5.1	3.0	1.7	4.2
Flag of this British nation					

History box

1801: Ireland becomes part of 'The United Kingdom of Great Britain and Ireland'.

1922: the Republic of Ireland gains independence.

1171: King Henry II of England takes control of Ireland.

1100: England, Scotland, Wales and Ireland are separate countries.

1276: King Edward I of England takes control of Wales.

1536: Henry VIII unites England and Wales.

1707: England, Scotland and Wales become 'Great Britain'.

Today: England, Scotland, Wales and Northern Ireland are still united as the UK.

Your turn

1 So – what about you?
- **a** Which country of the British Isles do you live in?
- **b** Which nation do you live in?
- **c** Which region do you live in? If you don't know, compare the maps on pages 64 and 139, and see if that helps.

2 This shows where Walter lives:

Show where you live, in the same way. (But if you live in Liverpool, do it for someone in Land's End. See the map on page 139.)

3 a Make a larger copy of the table on the right. (Just sketch the maps roughly, and fast.)
- **b** On your copy, colour in part of each map, to match its label. (So in the first one, colour in just Great Britain.)
- **c** Work out the population and area of the coloured parts, using the data at the top of the page.
- **d** Now give your table a suitable title.

4 How did the British Isles end up as two countries? The History box above gives key events.
- **a** Draw a timeline from 1100 to today. You could use a whole page, turned sideways. (For an example of a timeline, see question **2** on page 69.)
- **b** On your timeline, mark in arrows at the key dates, and add labels to show the events.
- **c** You can add small maps or flags or other symbols to your timeline. Give it a suitable title.

	Great Britain	United Kingdom	British Isles
Population (millions)			
Area (_____)			

5.3 What's our weather like?

Here you'll learn about weather patterns across the UK.

What is weather?

Weather means the state of the atmosphere. Is it warm? wet? windy?

Look at this weather map, for a day in October. Using the key, you can say that around **A** that day:

- it was quite cloudy and wet, but there was some sunshine.
- the temperature was around 6 °C.
- there was a south west wind (it blew *from* the south west).
- the wind was quite strong (around 38 miles per hour).

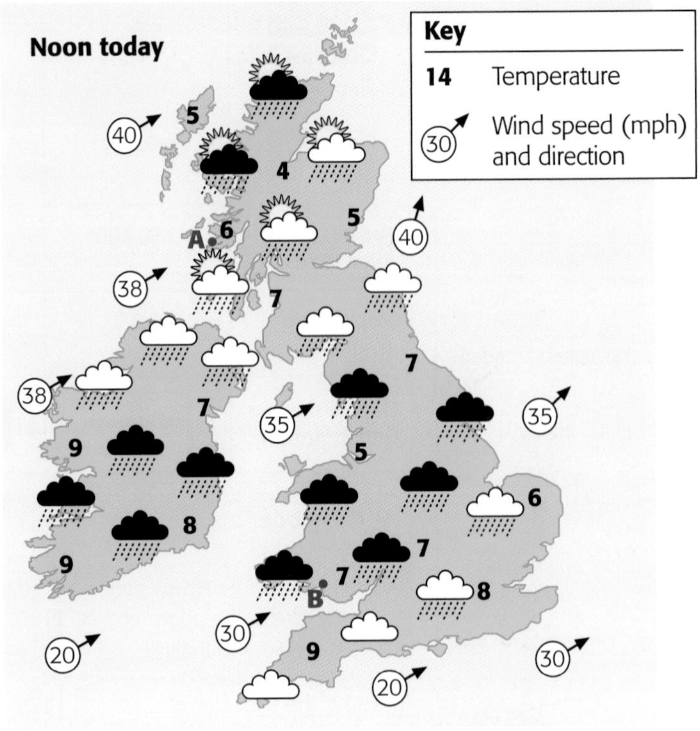

Noon today

Key
14 Temperature
(30) Wind speed (mph) and direction

Our weather is changeable

There are two key points about our weather:

- It can change from day to day.
- It can be different in different parts of the UK, on the same day. The weather map shows this.

What if ...
- ... it was warm and sunny every day?

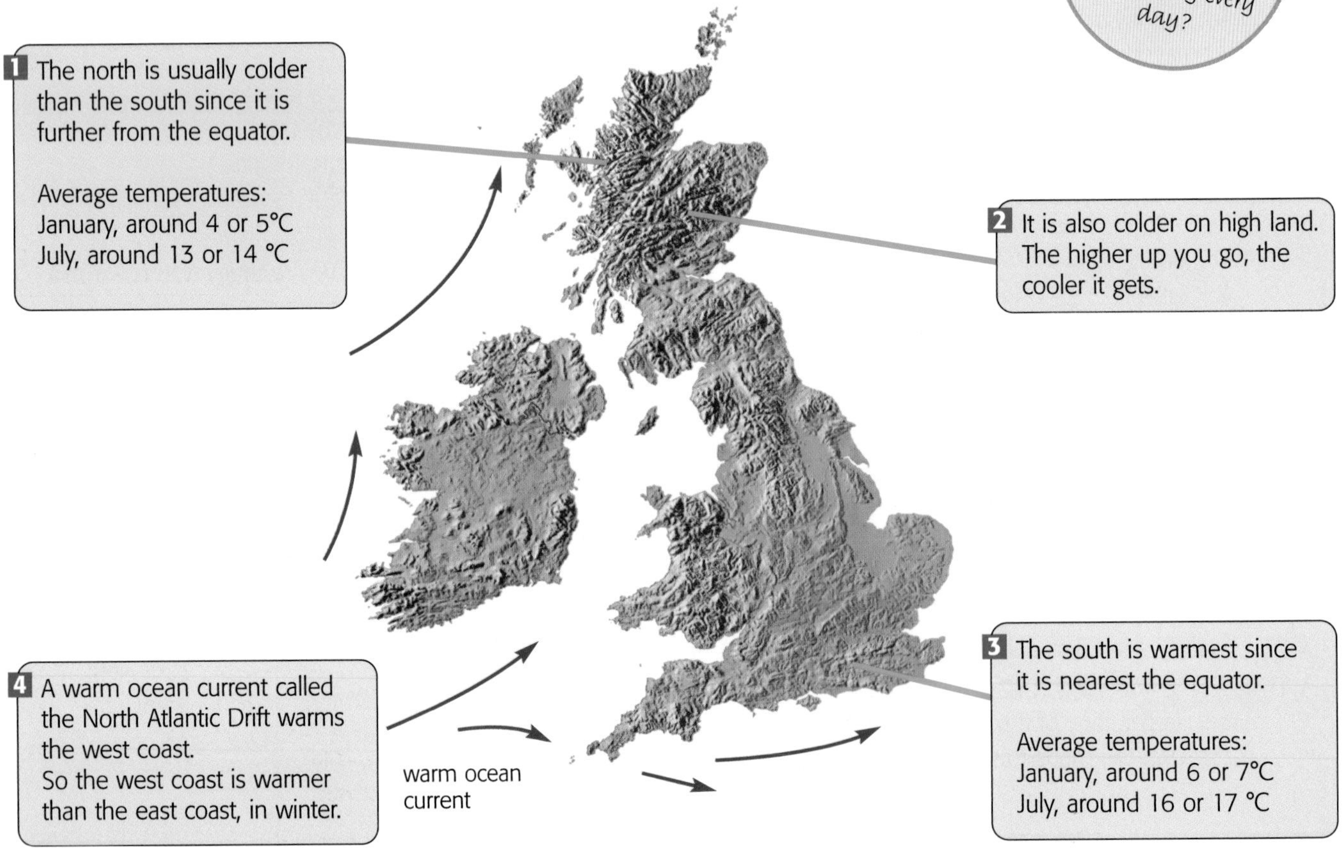

Which parts are colder? warmer?

Although the weather can change from day to day, there are some patterns. For example, some areas are usually colder than others.

1 The north is usually colder than the south since it is further from the equator.

Average temperatures:
January, around 4 or 5°C
July, around 13 or 14 °C

2 It is also colder on high land. The higher up you go, the cooler it gets.

3 The south is warmest since it is nearest the equator.

Average temperatures:
January, around 6 or 7°C
July, around 16 or 17 °C

4 A warm ocean current called the North Atlantic Drift warms the west coast.
So the west coast is warmer than the east coast, in winter.

warm ocean current

Which parts are wettest?

Look at the map on the right. It shows the average rainfall in a year, for the British Isles. As you can see, some parts get a lot more rain than others.

Overall, the higher parts are wetter. This is why:

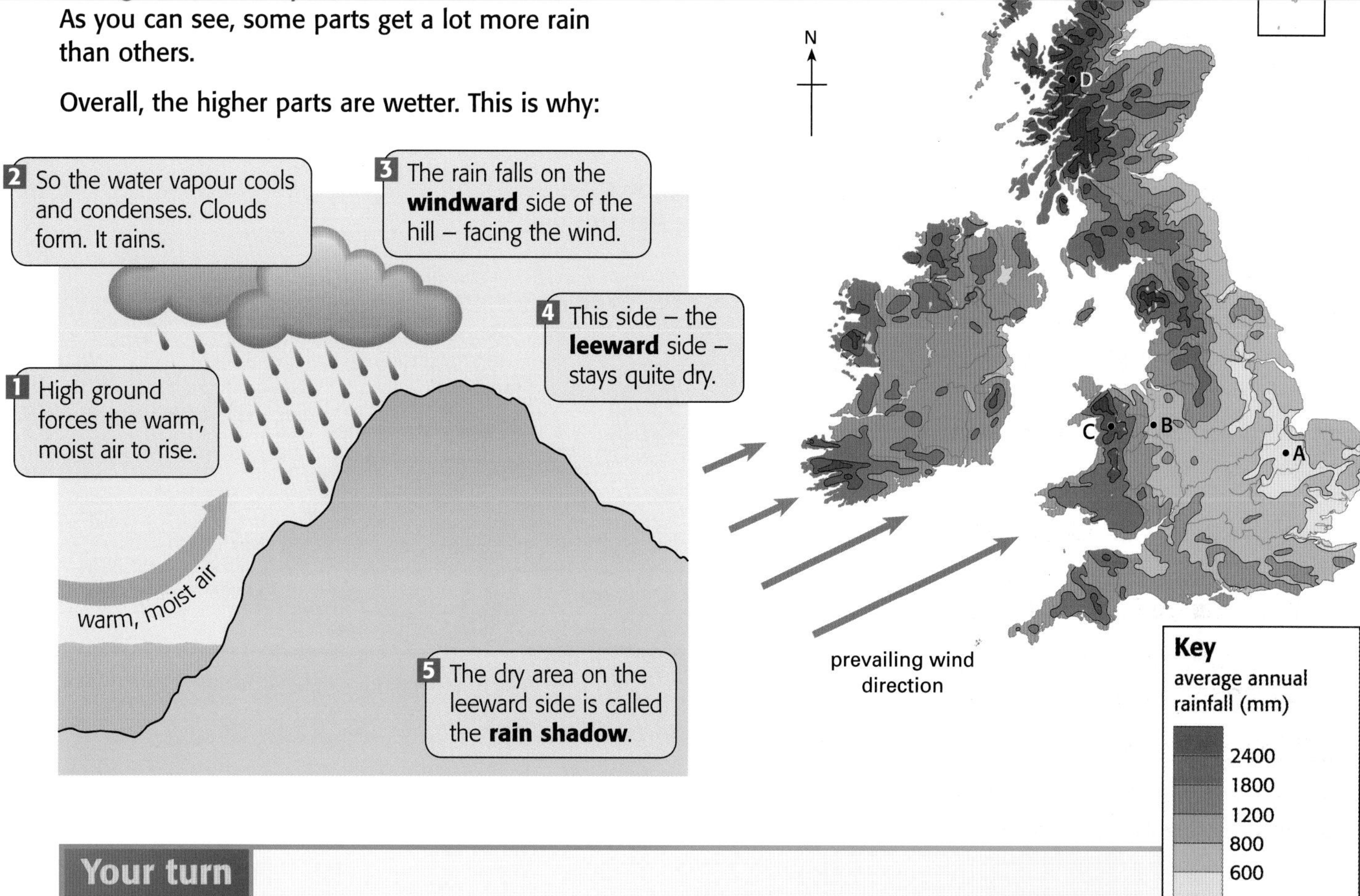

Your turn

1 What's the weather like where you are today? Describe it. You might be able to use some of these words:

sunny cloudy rainy dry calm
cold warm mild windy stormy

2 Look at the weather map at the top of page 66. Find the place marked **B**. Say what the weather was like around **B** that day, as fully as you can.

3 Look at boxes 1 – 4 on page 66. Does the weather map match what the boxes say? Give evidence!

4 Look at the rainfall map above. Four places are marked on it: **A**, **B**, **C** and **D**.
  - a Which one of them is the wettest?
  - b Which is driest?
  - c Which one has an average annual rainfall of:
    i 2000 mm? ii 500 mm?

5 Mountains help rain to form. How do they do this?

6 a What are *prevailing winds*? (Glossary.)
  - b The prevailing winds in the UK carry lots of moisture. Why? (Think about where they come from. The map on pages 140 –141 will help.)

7 a Overall, which side of Great Britain is wettest? See if you can explain why. (Page 62?)
  - b On the map above, **B** gets far less rain than **C**. Why?

8 And now, a challenge. On this map, the British Isles is divided into four zones.
  - a Make a larger, simpler, copy of the map.
  - b Colour the land in each zone in a different colour.
  - c Then add these four labels to their correct zones:

What's our weather usually like?

N

warm summers, mild winters, not so wet

mild summers, mild winters, wet

warm summers, cold winters, dry

mild summers, cold winters, not so wet

Who are we?

In this unit you'll find out how Britain has been peopled by immigrants.

The long march

An **immigrant** is a person who moves here from another country, to live. 10 000 years ago, nobody lived here. So all British people are descended from immigrants – even the Queen!

People from all over world have moved here. This drawing shows the main groups. There were many others, and new groups are still arriving.

Did you know?

- Before 1962, it was quite easy for people to move to the UK.
- Then a law was passed to make it a lot harder.

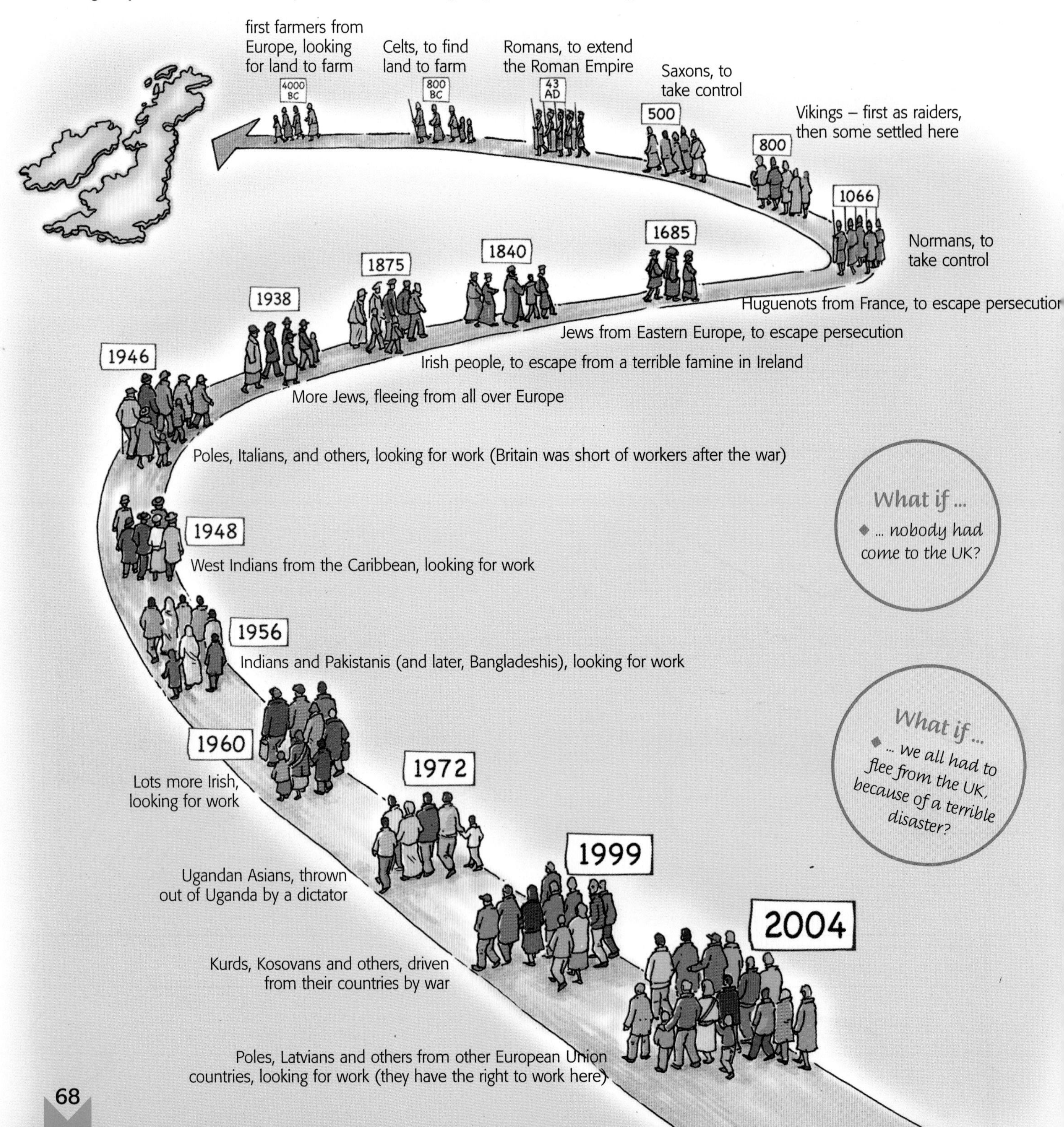

What if ...

- ... nobody had come to the UK?

What if ...

- ... we all had to flee from the UK, because of a terrible disaster?

All mixed up

So we carry the genes of past immigrants in our body cells. Exciting! Look at these.

And who are you descended from?

Your turn

1 What is an *immigrant*?

2 This shows the start of a **time line** for the main groups of immigrants since the year 1 AD.

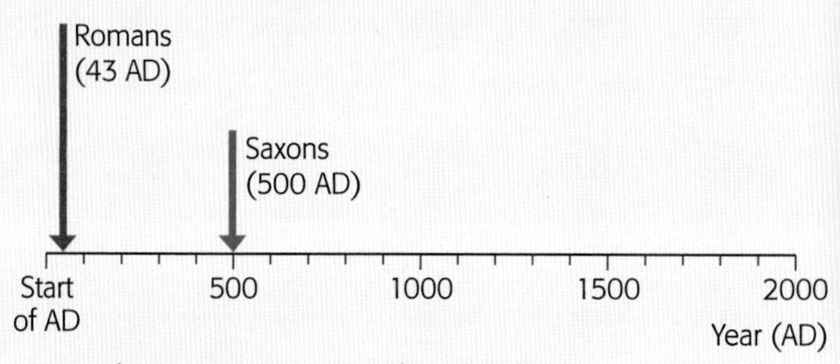

Now draw your own time line for them.

- **a** First draw a line 21 cm long. Divide and label it, with 1 cm for each century.
- **b** Draw an arrow for the Romans at 43 AD. Label it.
- **c** Repeat for the other groups in the table. After 1900 it gets crowded – so take care. (Try making all your arrows different lengths.)

3 Now look at your time line.

- **a** When was the biggest gap between new arrivals?
- **b** In which century did most new groups arrive?

4 Look at these different terms:

A refugee	**B** invader	**C** economic migrant
D emigrant	**F** settler	**E** asylum seeker

- **a** First, write down what each term means. (Glossary.)
- **b** Then choose what you think is the best term for each person in the photos on the right.

5 **a** Why did the very first immigrants come here?

- **b** What do you think attracts new immigrants today? Think of as many things as you can.

▲ *William the Conqueror, the Norman who took control of England by force in 1066.*

▲ *Chiyo in a Red Cross camp after her home was destroyed by an earthquake.*

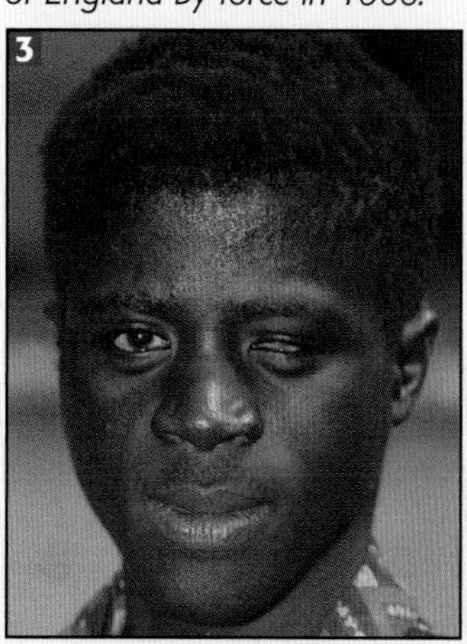

▲ *Philip, tortured by the army in his own country, has asked permission to stay here.*

▲ *Joy arriving from Jamaica in 1956. She wants to find a job.*

Where do we live?

Here you'll see how we humans have changed the UK, through where we chose to live.

Population density

About 64.5 million people live in the British Isles. About 60 million of them live in the UK. So are we all spread out evenly? Of course not.

The **population density** of a place is the average number of people per square kilometre.

The map below shows how population density changes around the British Isles. The deep green regions are the least crowded. The deep red regions are the most crowded.

The UK's 10 largest cities

	Name	Population (millions)
1	London	7.17
2	Birmingham	0.98
3	Leeds	0.72
4	Glasgow	0.58
5	Sheffield	0.51
6	Bradford	0.47
7	Edinburgh	0.45
8	Liverpool	0.44
9	Manchester	0.44
10	Bristol	0.38

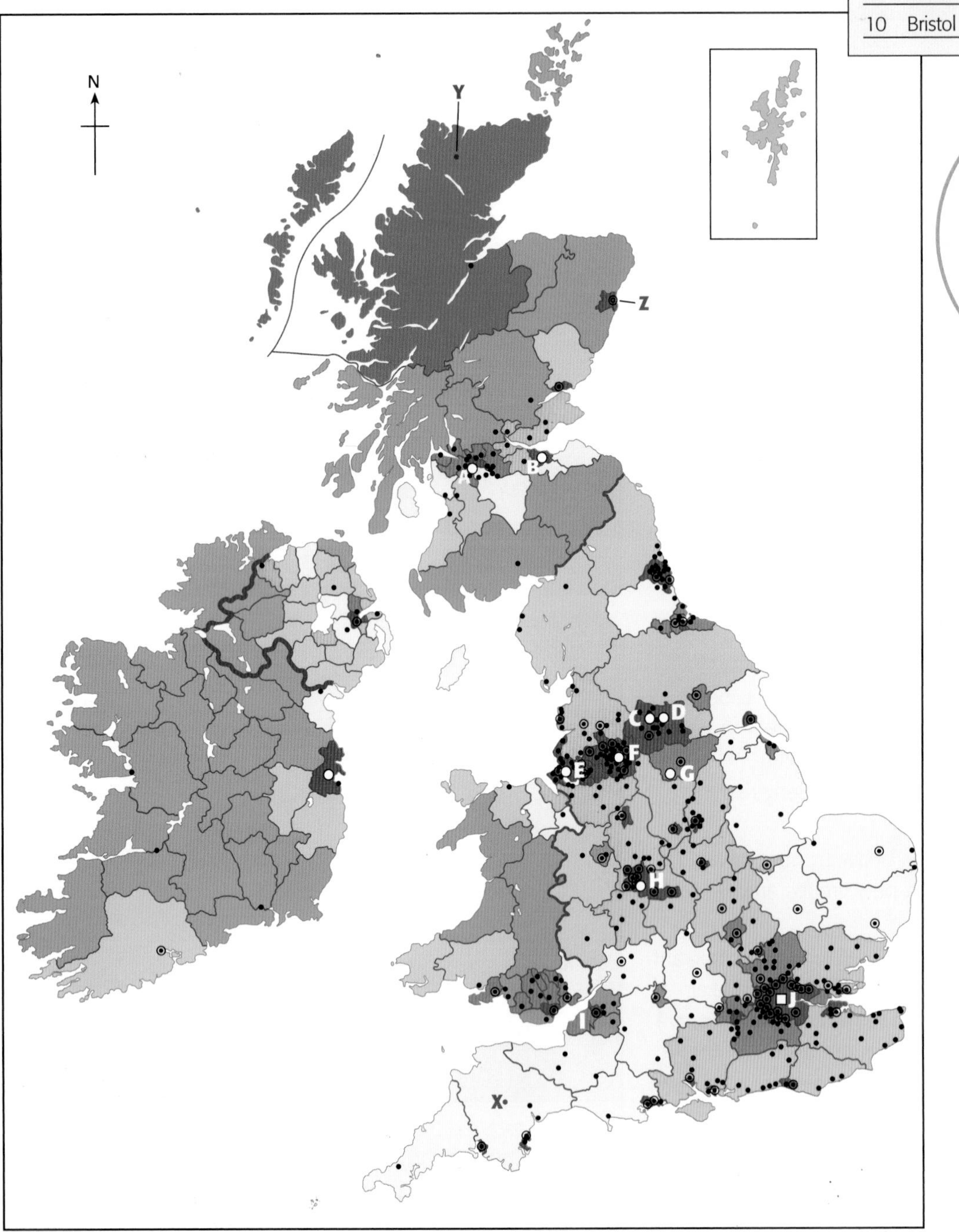

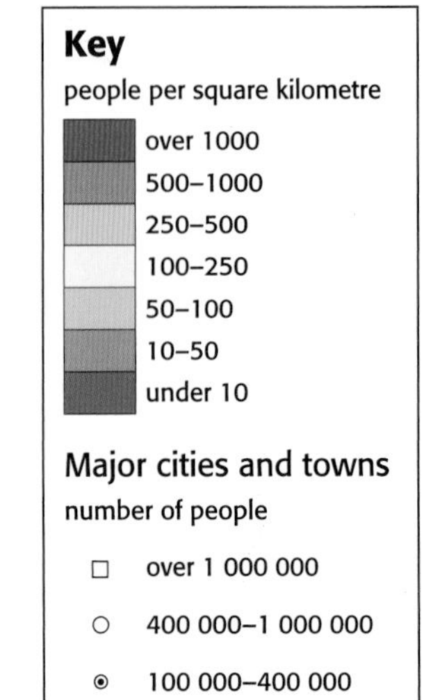

Some places in the UK are quite empty, while some ...

... are very crowded. What's your place like?

Look at photo A above. It shows a rural area – an area out in the countryside. Photo B shows a built-up or urban area. (It is London.)

Your turn

1 Look at the photos above. Which place has a higher population density, the urban area or the rural area?

2 See if you can name: **a** an urban area **b** a rural area near where you live.

3 Look at the map on page 70. It has letters marked on.

a What can you say about the population density:

i at **X**? ii at **Y**? iii at **Z**?

Use the term *people per square km* in your answer.

b Using the map on page 62, see if you can explain why the area around **Y** has such a low population density.

c i Overall, where are the two main areas of highest population density, in the British Isles?

ii See if you can explain why so many people live there. (Check out page 74?)

4 The *average* population density for the UK is 248 people per sq km.

Nation	Average pop. density (persons/sq km)
	387
	121
	66
	142

Copy the table above. Then, using just the map on page 70, see if you can fill in the names of the four nations (England, Wales, Northern Ireland, Scotland) in the correct places in the first column.

5 The table on page 70 shows the UK's top 10 cities. They are also marked on the map, with labels **A** to **J**. See if you can match each letter to the correct city. Start like this: **A** = ______ (Page 139 will help.)

6 Now look at this pie chart for the United Kingdom.

Where the UK population lives

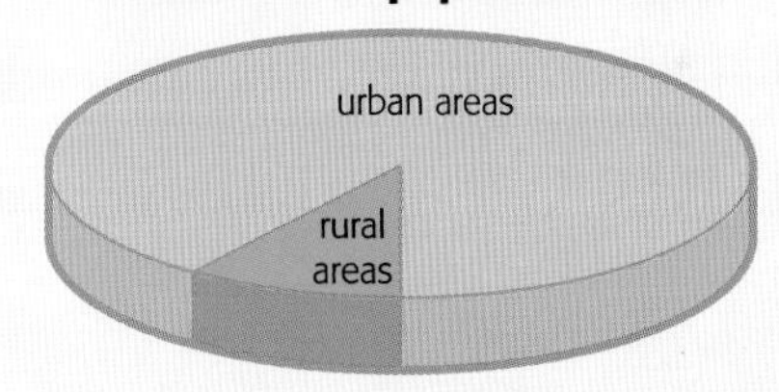

a Which statement is true?

i Most people in the UK live in the countryside.

ii About half of us live in towns and cities.

iii About $\frac{1}{10}$ of the UK population lives in rural areas.

b See if you can explain the pattern shown in the pie chart. (Why might people prefer to live in ____?)

7 Finally, use what you've learned to write a report called *The pattern of population density in the UK.*

- Make it at least 40 words long. (Try for more!)
- Say where the most and least crowded regions are.
- Try to include all these terms in your report:

highest land	*flat land*	*south east*	*England*
central	*coast*	*least populated*	*north*
Scotland	*Wales*	*Northern Ireland*	

- See what other relevant facts you can add.

What kind of work do we do?

Here you'll learn about the kinds of work people in the UK do for a living.

Different kinds of work

Economic activity is any work people get paid for.
(So homework does not count!)
You can divide it into four groups or **sectors**:

1 Primary
You gather materials from the Earth. For example grow crops, or dig for coal, or fish in the ocean.

2 Secondary
You make or build or process things, usually in factories. Like cars, clothing, frozen pizzas.

3 Tertiary
You provide a service. Like teach people, or look after them when they're ill, or carry them in a taxi.

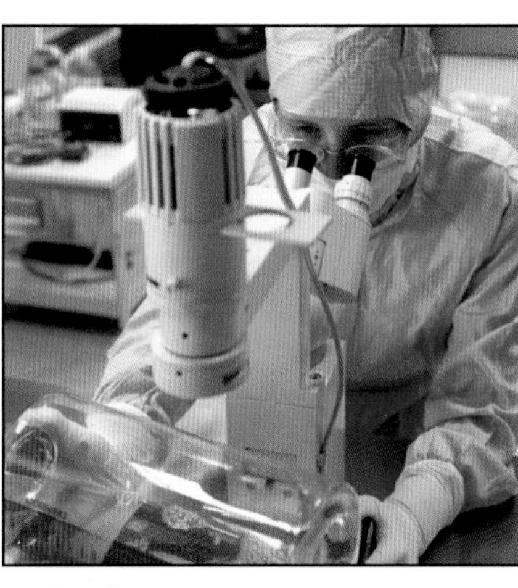

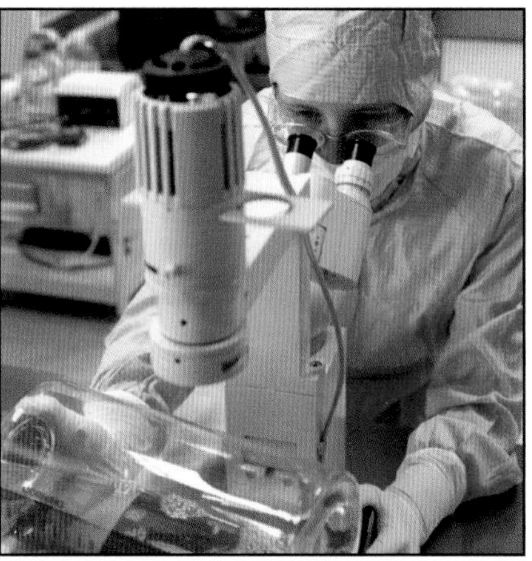

4 Quaternary
You do high-tech research to develop new things. Like new medicines or a new type of phone.

What if ...
◆ ... school paid you to do your homework?

What if ...
◆ ... everyone in the UK stopped working?

Industry

The word **industry** means a branch of economic activity.
The car industry is made up of all the companies that make cars.

At work in the UK

Altogether, about 26 million people in the UK work for a living. As this pie chart shows, most of them provide services. The number in the quaternary sector is too small to show up – but it is a very important sector.

What kind of work do we do, in the UK?

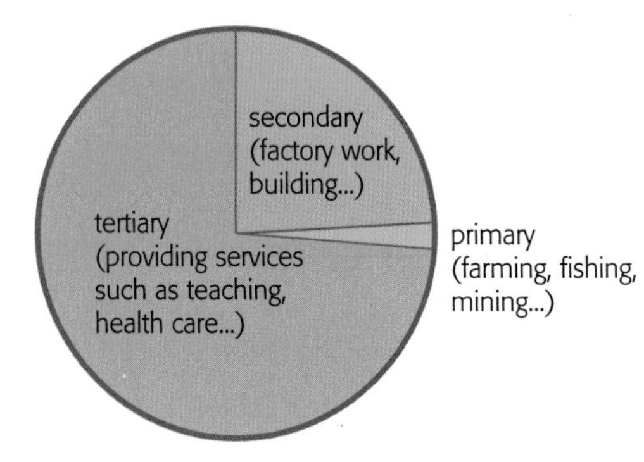

The changing pattern

Today, most of us provide services. But it was not always so.

500 years ago, the primary sector was largest. (Most of us were farmers.) 100 years ago, the secondary sector was largest. (Lots of people worked in factories.) This graph shows how the pattern has changed over time.

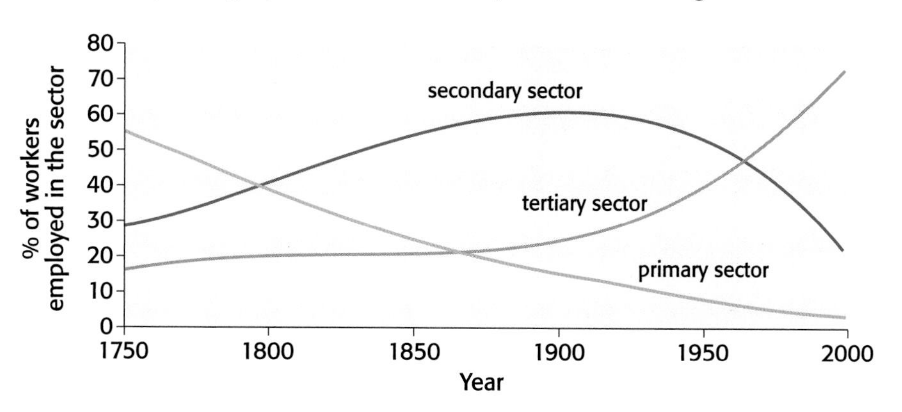

It is still changing. For example every year, more people leave farming. Every year, more factories close. They close because they can't compete with factories in countries like China, where wages are far lower.

Did you know?

- Tourists spend more than £35 billion a year in the UK.

What if ...

- ... there were no more farmers left in the UK?

Your turn

1 What is *economic activity*?

2 Does this count as economic activity? Give a reason.
- **a** going to school
- **b** doing a paper round
- **c** babysitting
- **d** tidying your room

3 a Make a table with headings like this:

Primary	Secondary	Tertiary	Quaternary

b Now write these jobs in the correct columns in your table:

nurse, postman, factory worker, oil rig worker, actor, gene researcher, football star, fireman, farmer, miner, architect, policeman, bank manager, house builder, inventor

c See if you can add *at least* ten other jobs to your table. Try for some for each sector.

4 Copy and complete this paragraph, using words from the brackets. The pie chart on page 72 will help.

In the UK most people earn a living by providing _____. The ______ sector employs about three times as many people as the ______ sector, and about _____ times as many as the ______ ______.

(*sector primary tertiary services secondary forty*)

5 Over 2 million people in the UK work in tourism. Tourists visit places. They travel, eat, and sleep.
- **a** Give six jobs connected with tourism.
- **b** Which sector do these jobs belong to?

6 This question is about jobs in different locations.
- **a** Make a larger copy of the table started here. Use a full page. Keep your drawings simple.

Location of job / Sector	rural area	sea	mountains	city
primary	farming	fishing		
secondary				
tertiary				
quaternary				

- **b** For each location, see if you can fill in *one* job for each sector. (It's hard, for some.) Two are in already.
- **c** Look at any cells you left empty. Explain why you left them empty.

7 Now look at the graph at the top of the page.
- **a** The secondary sector grew from about 1750 to 1900. What do you think caused this? (Hint: the I_____ R_____. Try the glossary?)
- **b** That sector is shrinking now. Give a reason.
- **c** Which sector just keeps on growing?

High or low earnings ?

Average pay is higher in some parts of the UK than others. Where? And why? Find out in this unit.

What people earn

This map shows how much workers earn a week, on average, in different regions. Why does it vary so much?

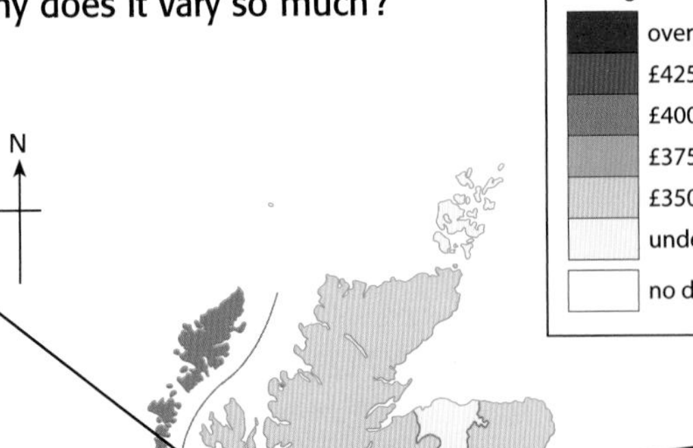
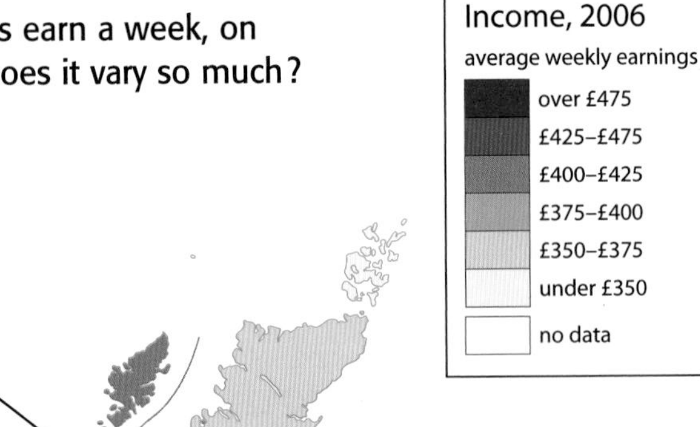

100 years ago this area was wealthy – thanks to shipbuilding and other heavy industries. But most of them died away and the area grew poor. Now it is recovering.

Earnings are high here because of the oil industry. The oil fields are in the North Sea.

This area was once wealthier because of coal. Now the mines are closed. (It's the same in Wales and other coal areas.)

This area has hardly any industry. But it has beautiful countryside and depends heavily on tourists.

Lots of hi-tech companies have set up in this area, around Cambridge. And there's good farmland.

This area never had much industry. The old tin mines are all closed. The land is not so good for farming. But tourists like it here.

Earnings are high in this area because it has lots of hi-tech companies (making computers, software, mobile phones and so on).

London is the capital city, and a major centre of government, business, and tourism. Companies have head offices here. There are many highly paid jobs.

This has some of the UK's best farmland. It grows large amounts of crops.

▲ *You can tell a lot about how wealthy or poor an area is, just by looking.*

Your turn

1 This list gives weekly earnings for five people. Work out their *average* weekly earning. (Hint: add up and divide by 5.)

Brian	£400
Liz	£50
Anna	£500
Joe	£250
Richard	£1000

2 Look at the map on page 74.
- **a** What's the average weekly earnings at A?
- **b** Does everyone at area A earn this much? Explain.
- **c** What's the average weekly earnings at B?
- **d** Give reasons for the big difference in the figures for A and B.

3 **a** Overall, where in Britain is average pay highest?
- **b** Where it is lowest?

Use terms like these in your answers: *south west, Wales, England, north of*. And page 139 may help.

4 The areas with highest average pay are the wealthiest areas. So what helps to make an area wealthy? Show your answer as a spider map, like this one:

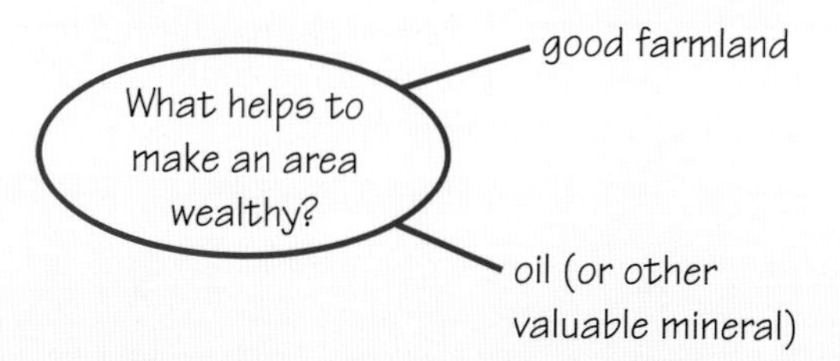

Page 74 will give you clues.

5 Now look at photos **X** and **Y** at the top of the page.
- **a** Compare the two places. Which one seems to be poorer? What is your evidence?
- **b** What clues can you find that this place is poorer than it used to be?
- **c** What might have caused this change?

6 The government tries to help poorer areas. For example by giving grants for:
- setting up new factories
- improving roads and tourist facilities.

This flow chart shows how a new factory can help a poorer area.

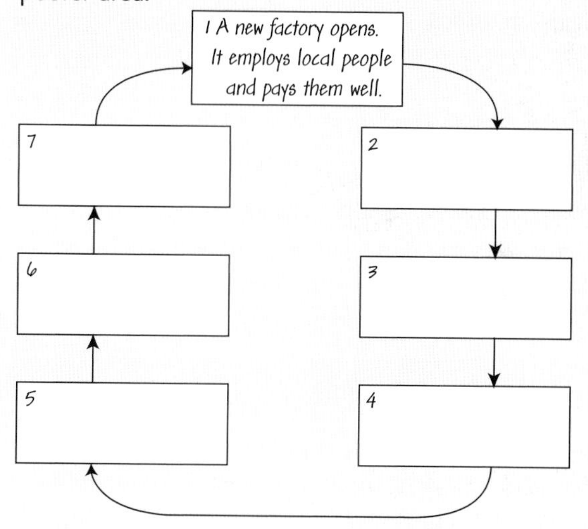

Make a much larger copy. (Use a whole page). Then write these in the correct boxes:

So the local people have more money to spend.

So the shops get better, and other services (like restaurants and sports centres) open.

So they buy more clothes and shoes and other goods.

... so more companies think about moving there.

So the area becomes more attractive to live and work in ...

So the local shops make more money.

7 Now do your own flow chart, to show how a new tourist attraction (for example a stunning new museum) could help a poor area.

The UK in the world

How is the UK linked to other countries? Find out in this unit.

What if ...
- ... the UK cut all its links with other countries?

The UK's political links

The UK has many political links with other countries. Look at this map:

The Commonwealth

- This is a group of 53 countries, including the UK.
- 51 of them were once part of the British Empire of colonies. (Mozambique is the exception.)
- The Queen is the Head of the Commonwealth.
- The Commonwealth countries have quite strong links with each other, including trade links.
- Since 1945, many of the immigrants to the UK have been from Commonwealth countries (such as India, Bangladesh and Pakistan).
- The Commonwealth Games are held every four years.

The European Union (EU)

- The EU is an organisation of European countries.
- By 2007 it had 27 members, including the UK. More countries hope to join.
- The EU countries trade freely together, with no barriers. They also co-operate in many other ways.
- If you belong to an EU country, you can visit other EU countries without a visa. You can also live and work in other EU countries.
- 15 of the EU countries use the same currency (money), the **euro**. But the UK uses pounds.

▲ *Many people from other Commonwealth countries have settled in the UK.*

▲ *Goods move freely between EU countries.*

Other links

The UK has many other links with the rest of the world.

1 Language and culture

For example:

- English is the language of many countries, and the main language used for business around the world.
- British football clubs have fans everywhere.
- Many British groups, and actors, and TV and radio programmes, are popular around the world.

2 Trade

The UK depends on trade with other countries.

- In 2005, it sold goods and services worth £322 billion to other countries. It bought goods and services worth £367 billion from other countries.
- It trades all over the world, but mainly with other EU countries.
- Outside the EU, its main trading partner is the USA. Inside the EU, it's Germany.

3 Tourism

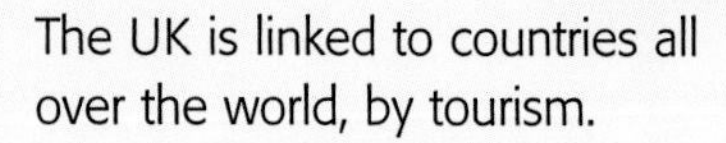

The UK is linked to countries all over the world, by tourism.

- There are over 30 million trips to the UK each year by people from other countries ...
- ... and around 70 million trips to other countries by UK residents.

REPUBLIC OF IRELAND

UNITED KINGDOM

4 TNCs

- Many British companies are transnational corporations (TNCs) with branches all over the world.
- Examples are Shell, Vodafone, and Barclays Bank.
- And many TNCs from other countries have set up branches in the UK. Like Sony, Coca-Cola, Nike, MacDonalds.

5 Treaties

TREATY

The UK has signed many treaties with other countries. For example:

- to reduce global warming
- to protect Antarctica
- against dumping waste at sea
- against whaling.

6 Aid

- The UK gives aid to poorer countries.
- In 2006 it gave almost £7 billion in aid.
- Most of this went to its ex-colonies, such as Ghana, Kenya, Zambia, and Uganda.

Your turn

1 The map on page 76 shows Commonwealth countries.
  a What is the Commonwealth?
  b Using the map on pages 140 – 141 to help you, name:
    i four Commonwealth countries in Africa
    ii three in Asia
    iii two in Oceania
    iv three in North America (include the Caribbean)
    v one in South America
    vi one in Europe
  c What do almost all Commonwealth countries have in common?

2 a Name six countries the UK has close links with, through the EU.
  b i Are *you* an EU citizen?
    ii How might this affect your life in the future?

3 a Do you think the UK *depends on* other countries?
  b Do you think other countries *depend on* the UK?
  Explain both your answers.

4 Which do *you* think are the two most important links with other countries, for the UK? Say why.

5 Now write a paragraph on *the UK's interdependence in the world.* (Glossary?) Not less than 60 words!

6 Rivers

The big picture

The Earth is always changing. It is changed by natural processes, and by humans. One thing that changes it naturally is the action of rivers.

This chapter is all about rivers. These are the big ideas behind the chapter:

- A river is just rainwater flowing to the sea.
- On the way it cuts and shapes the land, like a sculptor.
- It does this by picking up bits of rock and soil from one place, and dropping them in another.
- The result is special landforms along the river.
- We use rivers in many different ways, as they go on their journey. We harm them too.

Did you know?
- There are millions of rivers on the Earth.
- There are about 5000 in the UK. (Some are very small.)

What if ...
- ... our planet had no rivers?

Your goals for this chapter

By the end of this chapter you should be able to answer these questions:

- What is the water cycle?
- How does the rainfall from the water cycle feed a river?
- What do these river terms mean?

  source mouth channel bed banks tributary confluence drainage basin watershed flood plain
- How do rivers shape the land?
- How do these get formed?

  a V-shaped valley a waterfall a meander an oxbow lake
- In what kinds of ways do we use rivers? (At least five.)
- In what kinds of ways do we harm rivers? (At least three.)

Did you know?
- Every year the world's rivers carry about 10 billion tonnes of material to the sea.

What if ...
- ... you owned a river?

And then ...

When you finish the chapter, come back to this page and see if you have met your goals!

Your chapter starter

Look at the photo on page 78.

What are the people doing? Do you think it's dangerous?

Where did the water come from? And where is it going to?

Why is it swirling and foaming?

Do you think the river will look like this all the way along?

6.1

The water cycle

In this unit you'll learn about the water cycle, and how the rainfall reaches a river.

What is the water cycle?

Water sloshing around in the ocean this week may fall on you next week – as rain. It's the **water cycle** at work. Follow the numbers to see how water cycles between the ocean, the air and the land:

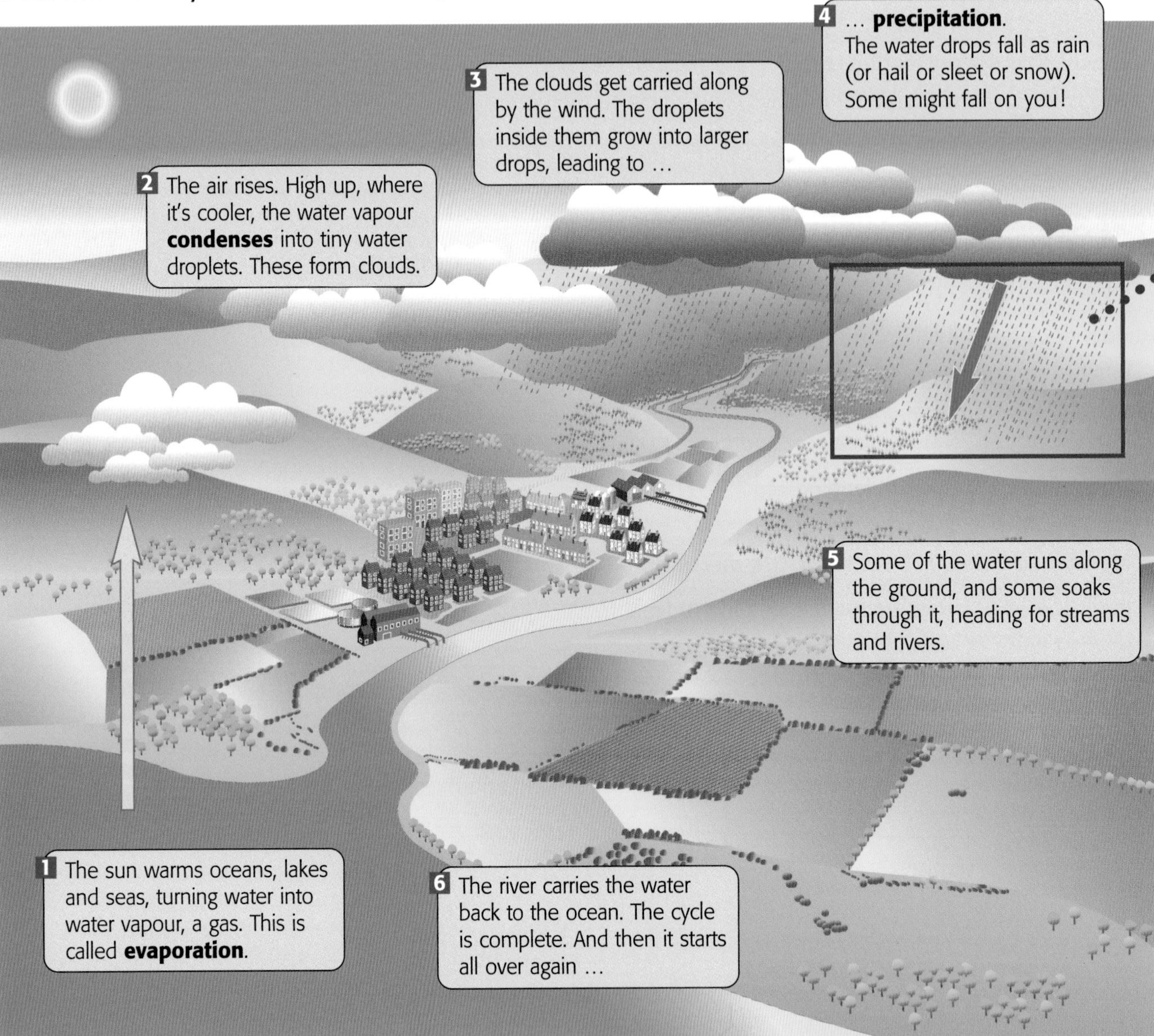

Did you know?
- The rain that falls on you has fallen millions of times before.
- It may have fallen on a dinosaur.

The water cycle and us

Without the water cycle we'd be in big trouble. We depend on rivers for water for homes and factories, and for spraying crops. And rivers depend on rain!

Every day, the UK 'borrows' about 47 billion litres of water from the water cycle. It is pumped from rivers and lakes and underground rocks, where rain has trickled down. It is cleaned up and used in homes and factories. And then our dirty water goes down the plug hole, is cleaned up in sewage works, and goes back to the river.

▲ *Borrowing from the water cycle.*

How rainfall reaches the river

Follow the numbers in order, to see the ways rain gets to a river:

1 If the ground is hard, or very wet, rainwater just runs along it. This is called **surface runoff**.

2 Otherwise the rain soaks into the ground. This is called **infiltration**.

This rock is **permeable**: it lets water seep through.

3 Some of the rain that soaks into the ground flows sideways through the soil. This is called **throughflow**.

4 The rest soaks down, and fills up the pores and cracks in the rock. Now it is called **groundwater**.

This rock is now soaking in groundwater.

5 Groundwater is always on the move. It flows along slowly.

6 A mixture of surface runoff, throughflow and groundwater feeds the river.

This rock is **impermeable**. It will not let water pass through.

Did you know?

- Right now, you are sitting above rock that's soaking in water.
- It's called groundwater.

Your turn

1 a Make a larger copy of this flow chart for the water cycle. (At least twice as large.)

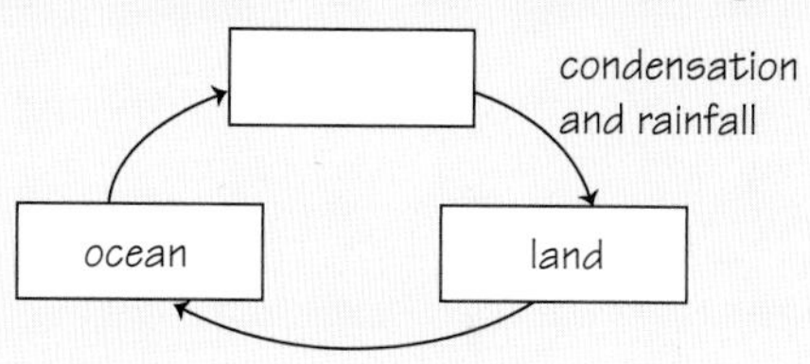

b Then add these labels in the correct places:
atmosphere *rainwater feeds rivers* *evaporation*

2 A – H below are definitions.

a You have to find the matching words, in the text!

b Then write out the words and their definitions.

A lets water pass through
B this water is held in rock, underground
C the name for water in gas form
D when water soaks down through the ground (*i....*)
E a longer name for rainfall
F the process that turns water into a gas (*e....*)
G the process that turns water gas into water
H does not let water pass through

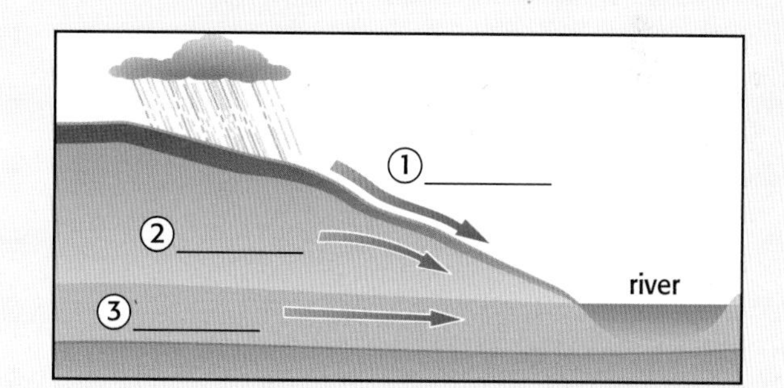

3 Draw a diagram like the one above, to show how rainwater reaches a river. Add the missing labels and a title.

4 Give reasons to explain why:

a rain does not sink down to the centre of the Earth
b a river can keep flowing even in very dry weather
c the river level falls, when there's a drought
d a river can fill up very fast in very wet weather

5 Suddenly the water cycle stops working. No more evaporation from the ocean! No more rain!
Write a radio report about the effect this will have on us. Make it dramatic – and not more than 250 words.

A river on its journey

In this unit you'll learn about the different parts of a river – and then take a look at the River Coquet.

> **Did you know?**
> ◆ There are around 5000 rivers in the British Isles.
> ◆ There are about 2 million in the USA.

The parts of a river

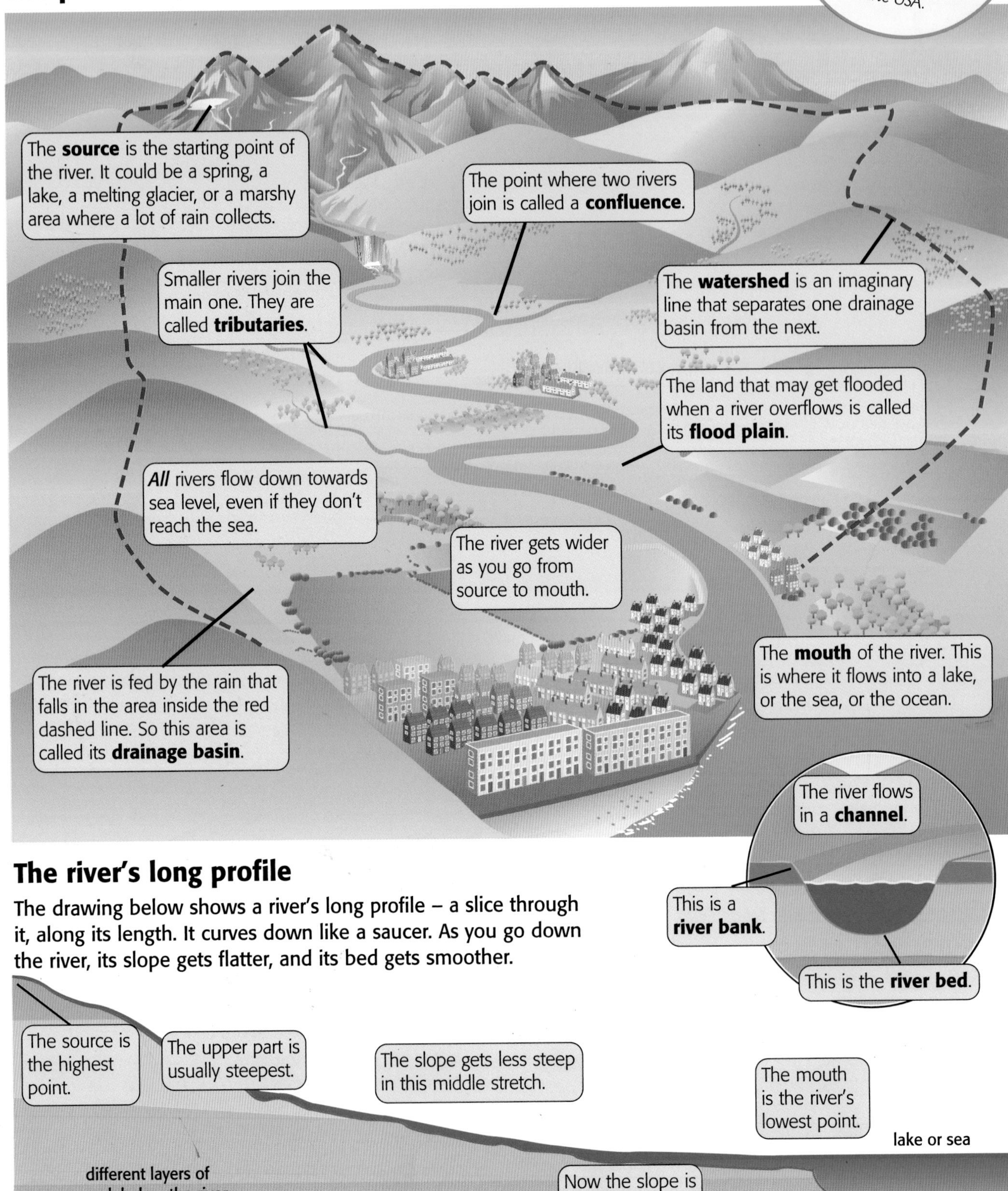

The river's long profile

The drawing below shows a river's long profile – a slice through it, along its length. It curves down like a saucer. As you go down the river, its slope gets flatter, and its bed gets smoother.

Your turn

The River Coquet and its drainage basin

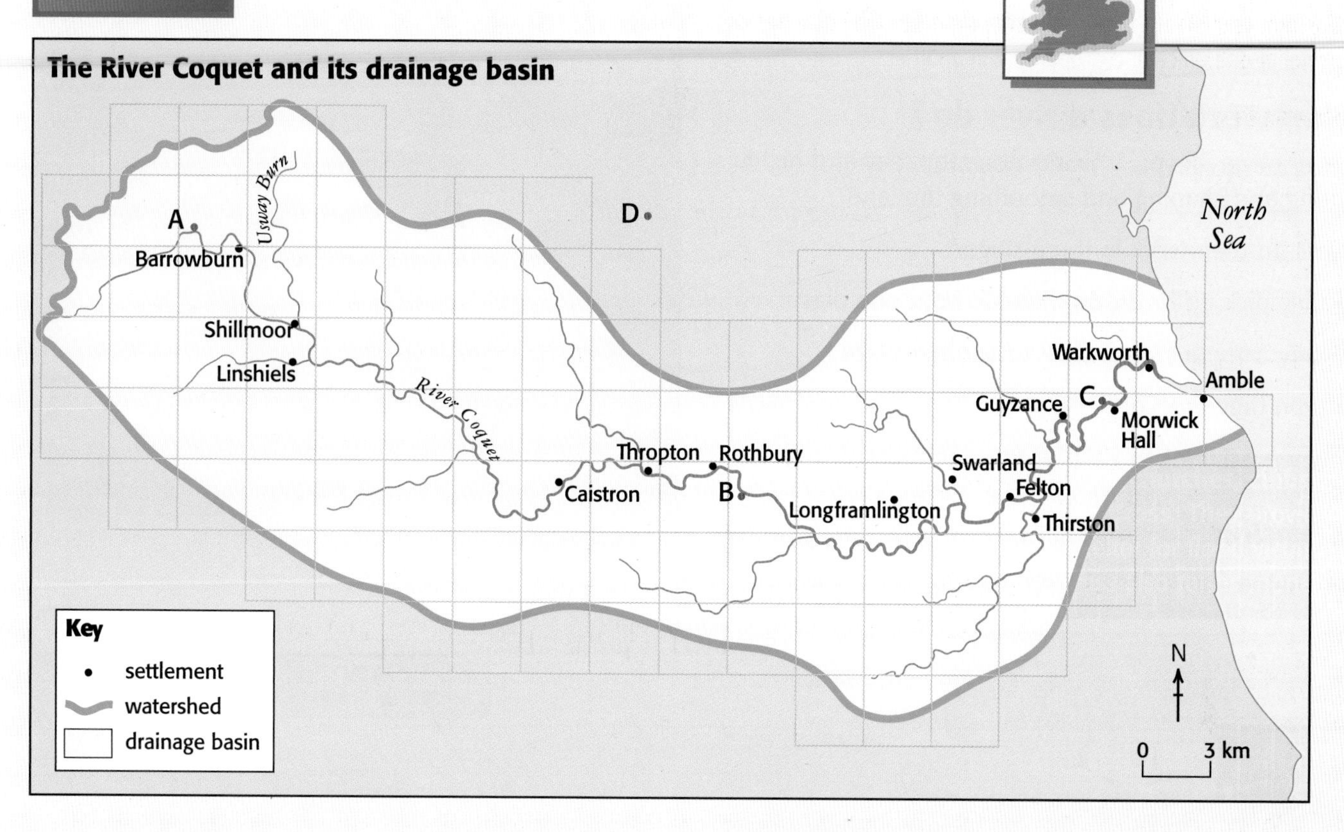

1 This is a map of the River Coquet. (You say *Cockette.*) It's in Northumberland.
 a Name the village nearest the source of the river.
 b How many tributaries join the river?
 c Name the settlement nearest the confluence of the Coquet and Usway Burn.
 d What sea does the Coquet flow into?
 e A, B and C mark three fields by the river. Which one:
  i is highest above sea level? ii is lowest?

2 a What is a *drainage basin*?
 b You can work out the area of the river's drainage basin roughly, by counting squares like this:

Full = 1.

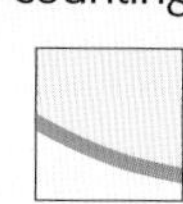

At least half full = 1.

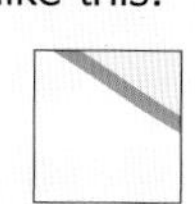

Less than half full = 0.

 i Count the yellow squares as shown.
 ii Each square represents 9 square kilometres. What is the area of the Coquet's drainage basin?

3 How long is the Coquet? Use the scale.
 a 55 km b 80 km c 110 km

4 Where is the river's long profile likely to be steeper?
 a between Barrowburn and Shillmoor
 b between Guyzance and Morwick Hall
 Explain your answer.

5 Rain falls at D on the map above. Will it end up in the Coquet? If not, where do you think it will go?

6

This photo was taken at the mouth of the Coquet. (The OS map on page 31 also shows this area.)
 a In which direction was the photographer facing?
 b Name the village on the right.
 c This village used to be a Roman settlement. Suggest reasons why the Romans chose this site.
 d What clues can you find from the photo, that this place is now:
  i a fishing port? ii a holiday resort?
 e You are planning a day out there. What will you do?

Rivers at work

In this unit you'll learn how rivers shape the land, by picking up material in one place and dropping it in another.

▲ *Hard at work ...*

What work does a river do?

A river never sleeps. It works non-stop, day and night, cutting and shaping and smoothing the land.

Rivers do their work in three stages:

1 they pick up or **erode** material from one place

2 they carry or **transport** it to another place

3 then they drop or **deposit** it.

Now we will look at each of these in more detail.

1 Erosion

This shows the different ways erosion takes place:

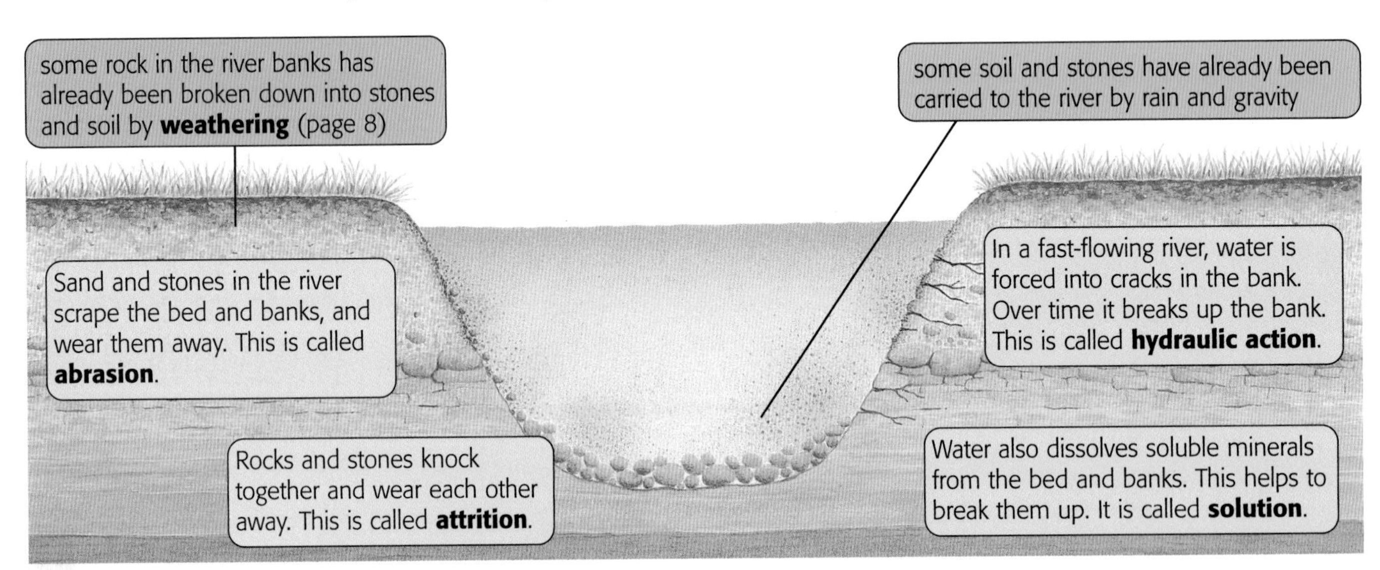

The faster it flows, and the more water it has, the faster the river erodes.

2 Transport

The material the river carries is called its **load**.

The faster it flows, and the more water it has, the larger the load the river can carry.

3 Deposition

When it reaches flatter land, the river slows down. It no longer has the energy to carry its load, so it deposits it – just like you put things down when you are tired. The deposited material is called **sediment**.

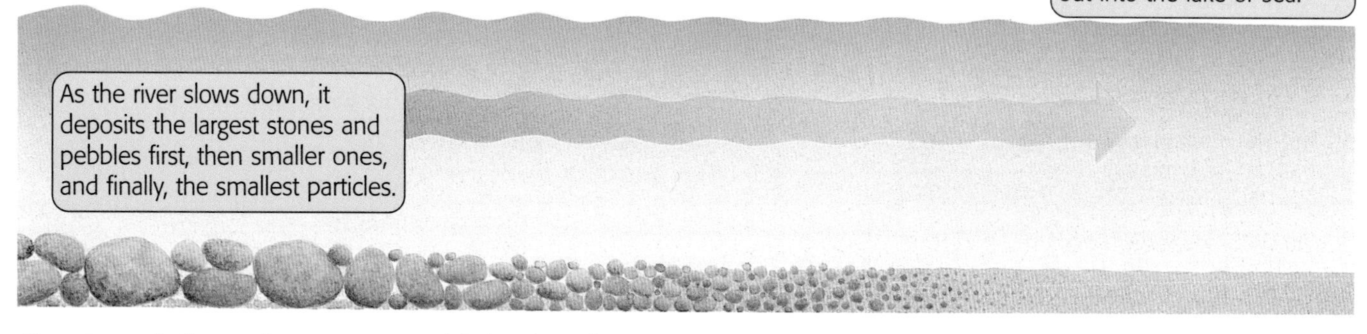

The slower it flows, the more material the river deposits.

Where does all this happen?

Your turn

1 **A** material is carried away / material is picked up / material is dropped — **B** erosion / deposition / transport

a List A shows the jobs that go on in a river. Write them in the correct order.

b Beside each, write the correct term from B.

2 Now look at the photo on page 84.

a What work do you think the river is doing there? Say why you think so.

b What part is being played in this work by:

i the water itself? ii stones in the river?

Give the correct name for each process you mention.

3 a Look at this photo. What job is the river doing at **X**?

b Do you think the river is flowing quickly, or slowly, in this stretch? Explain why you think so.

c Is this area in the flood plain? Explain your answer.

4 During a heavy flood a river can transport trees and large boulders. Explain why.

5 Look again at the photo on page 84. It was taken in March, after a lot of heavy rain.

a In what ways might the river be different at the end of a very dry summer? Give your reasons.

b How might this affect the work the river does?

6.4 Landforms created by the river

In this unit you'll learn about the landforms a river creates, by eroding and depositing material. They are shown in this drawing:

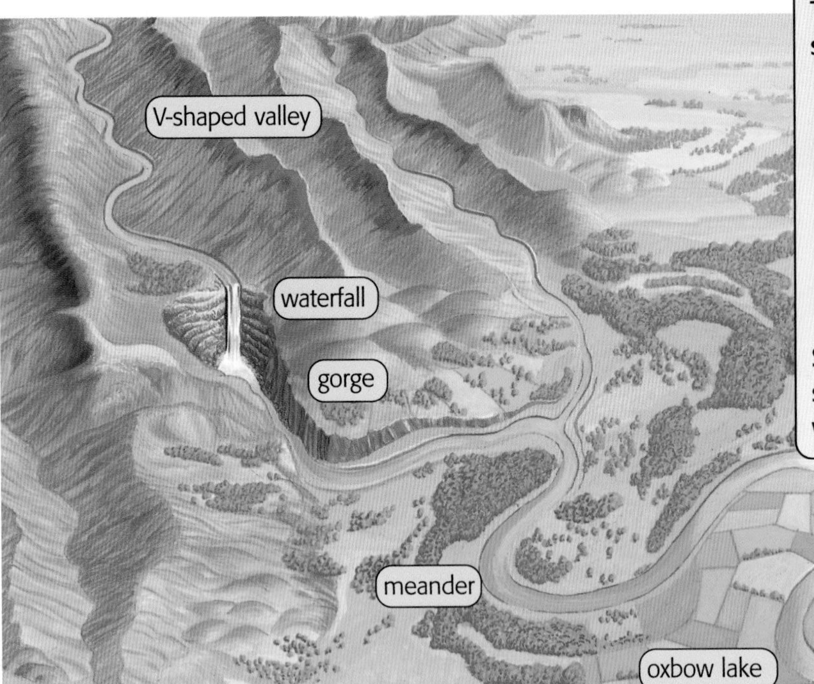

A V-shaped valley

The river cuts down through the land like a saw. This is called **downward erosion.**

Soil and stones then get carried down the slopes by rain and gravity – so the V grows wider and wider.

A waterfall

The water tumbles over a ledge of hard rock.

A waterfall is a sign that there are layers of different rock under the river.

The rock in the top layer does not erode easily. That's why the waterfall forms.

How a waterfall develops

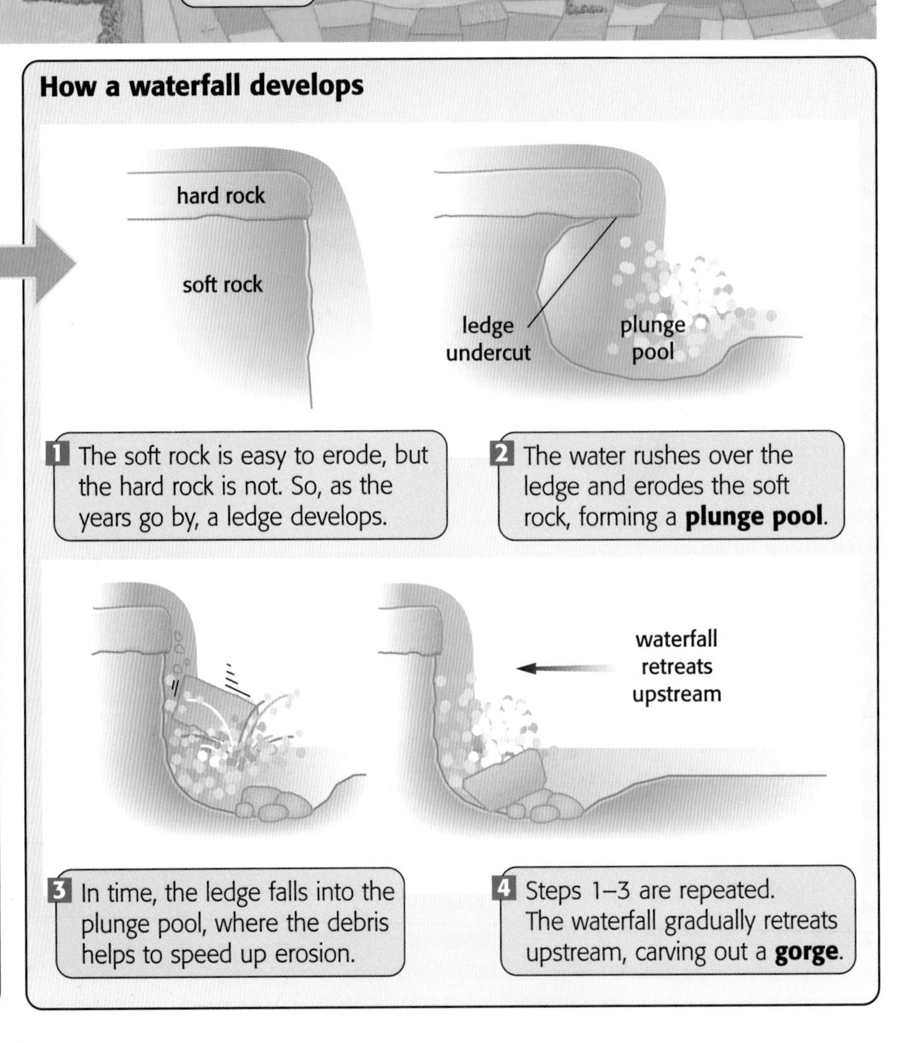

1. The soft rock is easy to erode, but the hard rock is not. So, as the years go by, a ledge develops.
2. The water rushes over the ledge and erodes the soft rock, forming a **plunge pool**.
3. In time, the ledge falls into the plunge pool, where the debris helps to speed up erosion.
4. Steps 1–3 are repeated. The waterfall gradually retreats upstream, carving out a **gorge**.

How a meander develops

A meander starts as a slight bend. Look how it develops:

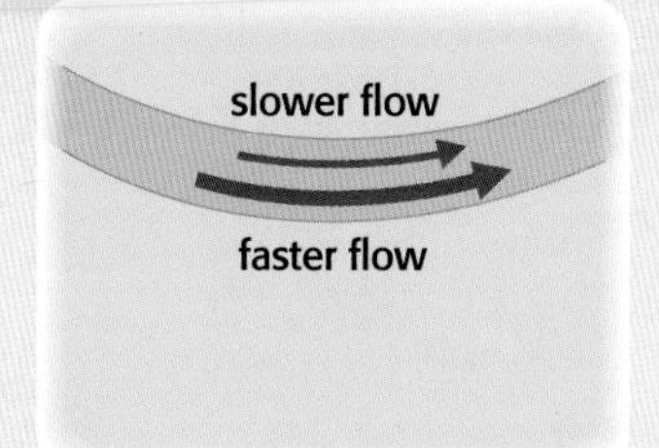

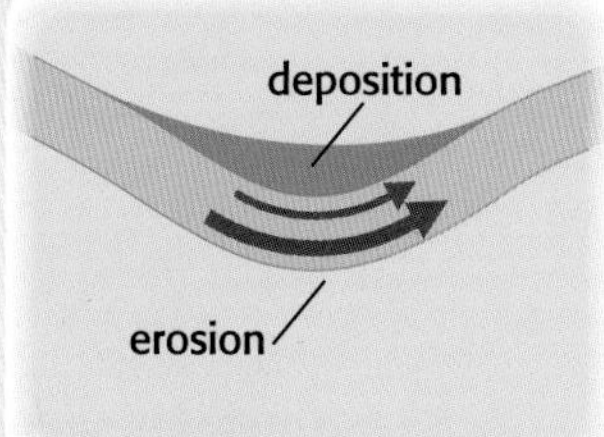

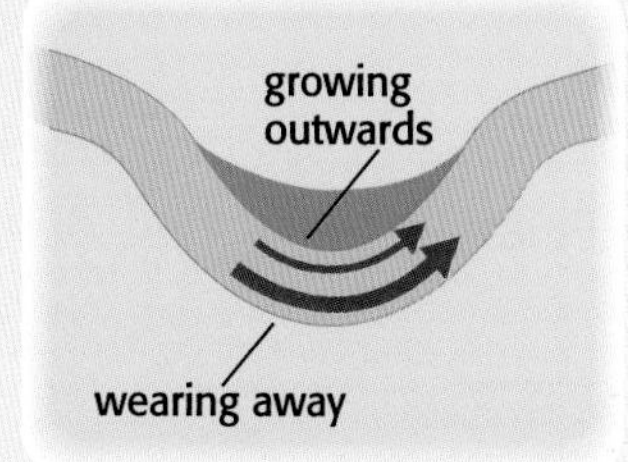

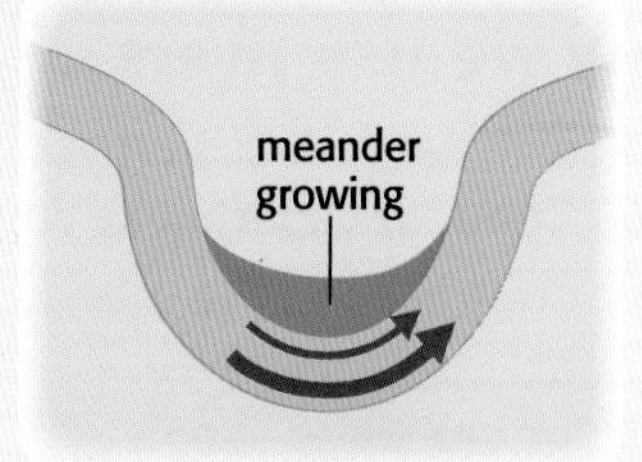

1 Water flows faster on the outer curve of the bend, and slower on the inner curve. So …

2 … the outer bank gets eroded, but material is deposited at the inner bank. Over time …

3 … as the outer bank wears away, and the inner one grows, a meander forms.

4 As the process continues, the meander grows more 'loopy'.

How an oxbow lake develops

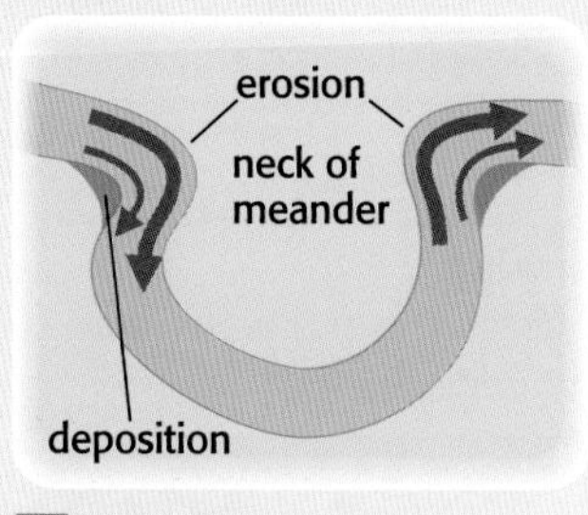

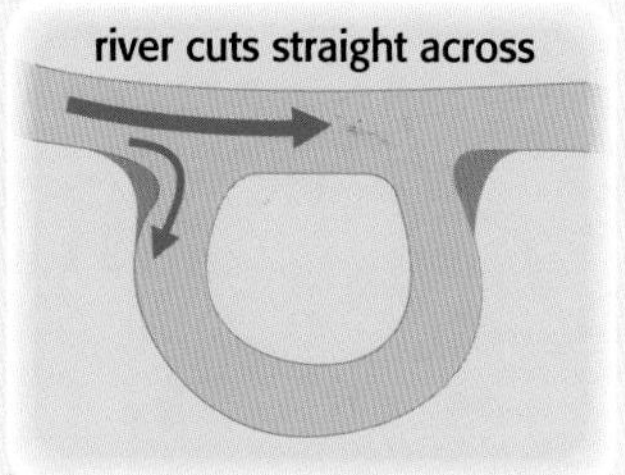

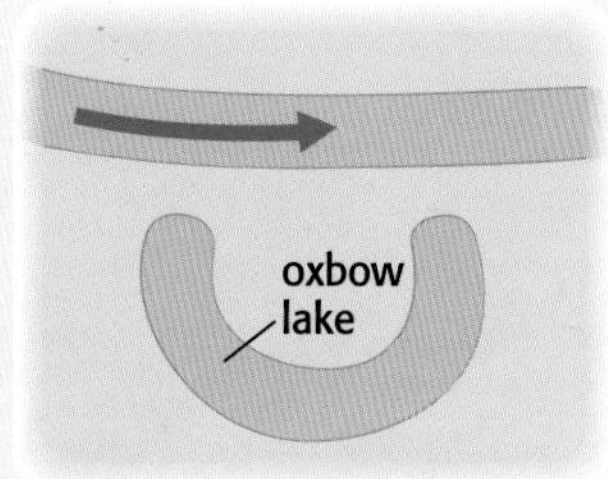

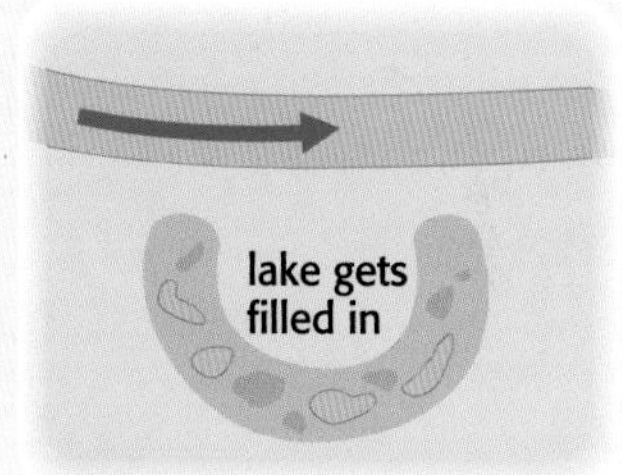

1 An oxbow lake starts as a big meander like this. Thanks to erosion and deposition …

2 … the neck grows narrower and narrower. And eventually the river just takes a shortcut.

3 Soon the loop is sealed off altogether. It turns into an oxbow lake.

4 In time the lake will get covered with weeds, and fill with soil, and disappear.

Your turn

1 Make a table like this one and complete it for all the river landforms shown on these two pages.

Landform	Created by …
V-shaped valley	erosion

2 What is a *gorge*? (Try the glossary.)

3 A river is flowing over layers of rock, like this:

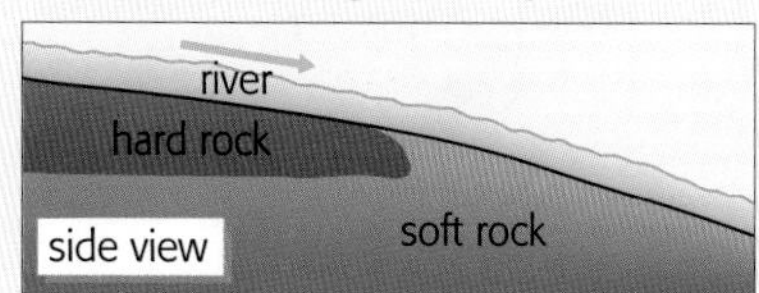

a Which will erode faster, the hard rock or the soft rock?

b Draw diagrams to show how a waterfall will eventually develop.

c Then show how a gorge will form.

4 Look at the scene above.

a What is happening at **A**? Why?

b What is happening at **B**? Why?

c Draw and label a sketch to show how this river might look 100 years from now.

Rivers and us

We make use of rivers, as they go on their journey. We harm them too.
In this unit you'll see how.

How we use rivers ...

Ever since we humans arrived on the Earth, we have used rivers. For water, food, transport, washing things, dumping stuff. And we still do.

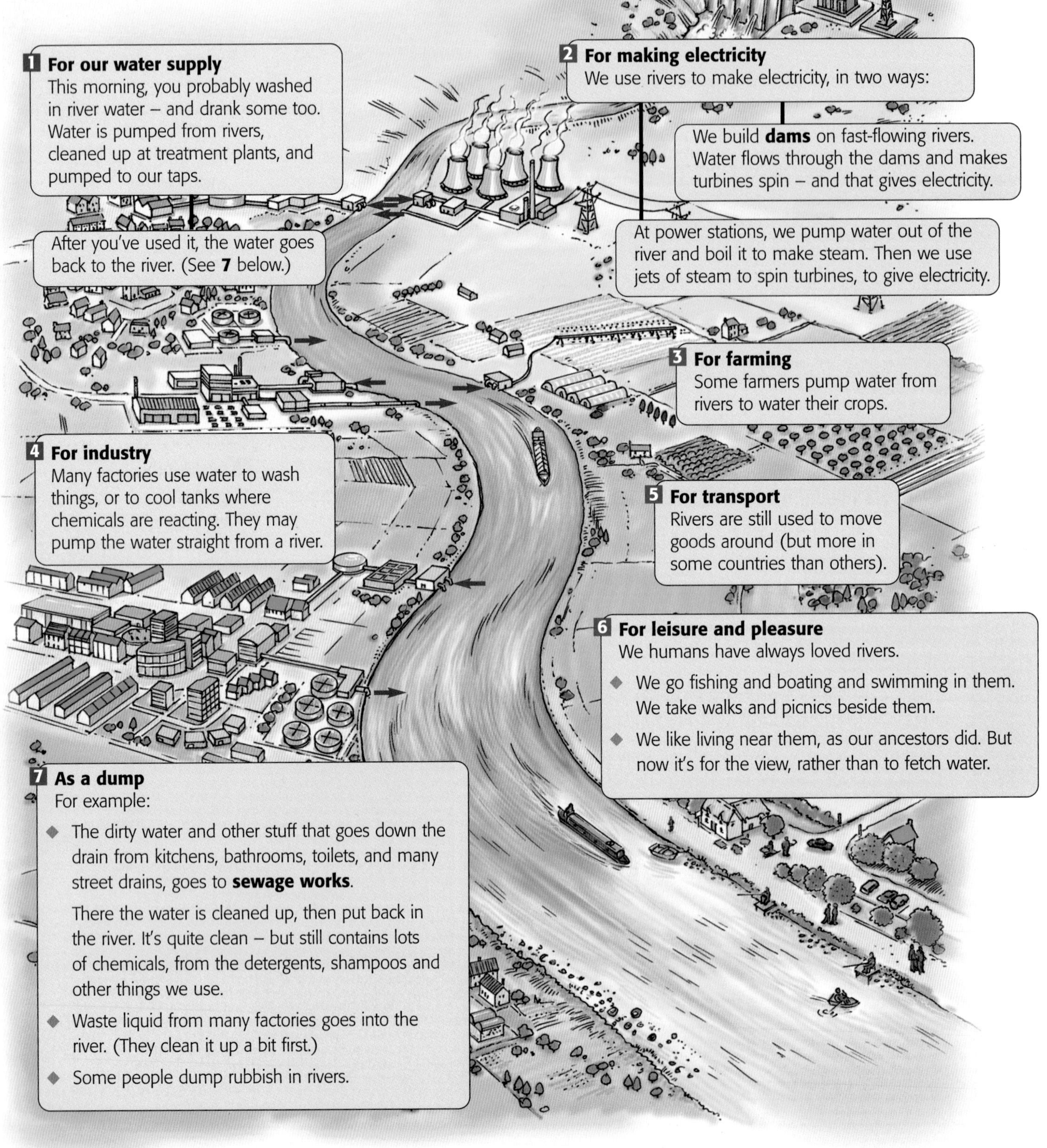

1 For our water supply
This morning, you probably washed in river water – and drank some too. Water is pumped from rivers, cleaned up at treatment plants, and pumped to our taps.

After you've used it, the water goes back to the river. (See **7** below.)

2 For making electricity
We use rivers to make electricity, in two ways:

We build **dams** on fast-flowing rivers. Water flows through the dams and makes turbines spin – and that gives electricity.

At power stations, we pump water out of the river and boil it to make steam. Then we use jets of steam to spin turbines, to give electricity.

3 For farming
Some farmers pump water from rivers to water their crops.

4 For industry
Many factories use water to wash things, or to cool tanks where chemicals are reacting. They may pump the water straight from a river.

5 For transport
Rivers are still used to move goods around (but more in some countries than others).

6 For leisure and pleasure
We humans have always loved rivers.

- We go fishing and boating and swimming in them. We take walks and picnics beside them.
- We like living near them, as our ancestors did. But now it's for the view, rather than to fetch water.

7 As a dump
For example:

- The dirty water and other stuff that goes down the drain from kitchens, bathrooms, toilets, and many street drains, goes to **sewage works**.

  There the water is cleaned up, then put back in the river. It's quite clean – but still contains lots of chemicals, from the detergents, shampoos and other things we use.
- Waste liquid from many factories goes into the river. (They clean it up a bit first.)
- Some people dump rubbish in rivers.

▲ *Having fun on the river.*

▲ *A river – after a leak from a factory making orange drinks.*

... and how we abuse them

Rivers are full of living things: fish, insects, plants, and tiny organisms. Their banks are home to ducks and other river birds. So how do we look after all these?

- We put poisonous things down drains, and they end up in the river. Things like oil, paint, chemicals from factories. All can kill.
- Waste liquid from homes and farms contains **nitrates** and **phosphates**. These chemicals help tiny plants called algae to grow like a carpet over a river, blocking out sunlight. When the algae die, bacteria feed on them, using up the oxygen dissolved in the water. Fish die without oxygen.
- Warm water from factories and power stations gets emptied into rivers. If it's too warm, it kills fish and other river life.
- Fish and river birds get tangled up in rubbish we dump in rivers. If they swallow bits of plastic bags, these fill their stomachs up. So they can't eat, and starve to death.

In the past, we almost ruined many rivers. Now we take more care. If people, or companies, are caught polluting rivers, they are fined. People even get sent to jail.

But the 'clean' water we put back in the river from sewage works still contains harmful chemicals. This is still a problem.

▲ *Killed by toxic chemicals.*

Did you know?

- *Some places pump their water up from* ***aquifers****.*
- *These are large areas of rock below ground, full of groundwater.*

Your turn

1 *'Rivers have nothing to do with me.'* True or false? Why?

2 a Show all the ways we use rivers. Give your answer as a spider map. You could start like this:

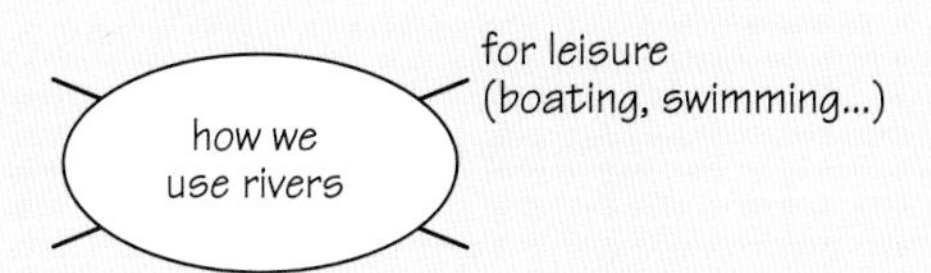

You can add little pictures if you like. (Work fast!)

b Now number them in what you think is their order of importance to us. (1=the most important.)

3 What if fish could speak? Make up a conversation between two fish, about humans. The fish live on the River Aire. (Look at the map on page 139.)

4 Do *you* harm rivers in any way? Explain.

5 Look at this person. Do you agree with the message on his placard? Give your reasons.

6 Come up with an idea for a poster to stop people putting harmful things (like waste engine oil) down drains.

7 Floods

The big picture

This chapter is about floods in the UK. These are the big ideas behind the chapter:

- Our planet holds many dangers for humans. One is floods.
- There have always been floods. But we are making them worse.
- Floods can do enormous damage.
- Experts say the UK will have more storms and floods in the future, thanks to global warming. We need to be prepared.
- There are several things we can do to control or prevent floods, and make homes more flood-proof.

Your goals for this chapter

By the end of this chapter you should be able to answer these questions:

- What do these terms mean?

  flooding *flood plain* *embankment* *infrastructure*
  short-term response *long-term response* *insurance*
- What's the main cause of floods?
- Which *natural* factors increase the risk of floods?
- In what ways have we humans increased the risk of floods?
- What are some consequences of flooding, after the water has gone …
  for the local area? for the rest of the country?
  (Give at least three examples for each.)
- What kinds of groups help out, in serious flooding? (Name at least four.)
- What sort of things can we do, to prevent floods? (Give at least four.)
- What can we do to make our homes more flood-proof? (Give at least three things.)

And then …

When you finish the chapter, come back to this page and see if you have met your goals!

Did you know?

- In the UK, 1 in 12 of us lives in an area at risk of flooding. (Do you?)

Did you know?

- China's Yellow River has the world's worst record for flood deaths.
- In 1887, 900 000 people drowned when it flooded …
- … and millions more died later from hunger and disease.

What if …

- … we had terrible floods in the UK every summer?

What if …

- … the government made us all move to high land?

Your chapter starter

Look at the photo on page 90. What's going on here?

How do you think the people feel about it?

What kinds of problems do you think it's causing?

Do you think it's anyone's fault?

What do you think will happen to all the water? And when?

Tewkesbury under water

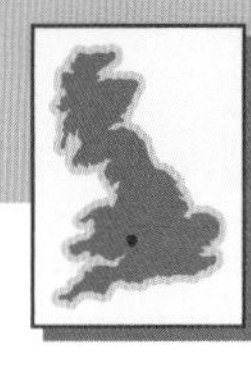

In Summer 2007, floods hit many parts of the UK. Here you can find out about the floods that hit Tewkesbury in Gloucestershire.

Some holiday!

Ellen lives in Tewkesbury. This is how her summer holiday started, in July 2007:

Friday 20 July

School over. They let us off at lunchtime. I get totally soaked on the way home. It has been raining for days. Now it's bucketing.

Dad gets home from work early. Looks very worried. Says the river is going to flood. By the time it's dark, the bottom of the road has turned into a lake. There are cars floating in it. I think one is Mr Lenon's.

Saturday 21 July

Doorbell wakes me. It's a man in a yellow jacket, giving out sandbags from his truck. The lake at the bottom of the road is getting closer. Dad looking even more worried.

Dad puts the sandbags in front of the door. Mum shouts 'Food for the forces' and goes off in her wellies. She comes back with torches and bags and bags of food. She says 'Now stop messing around and start carrying things upstairs.'

Lots of stuff upstairs now, except big things like the sofa. A lot is stuffed into my room, worse luck. Dad is in the attic looking for the camping gas stove. He sticks his head out the hatch and shouts 'Charge your mobile phones, you lot, and don't waste them chatting.' Okay,done.

He thumps up and down stairs all night, checking. Keeps us all awake.

Sunday 22 July

Get up early. Water coming in under the front door, in spite of the sandbags. Scary! Dad turns off the gas and electricity at the mains. He says 'We don't want an electric shock or a gas leak, do we?'

Spend the day upstairs, looking out the window. Street is like a river. No computer, no TV, nothing to do. We see all kinds of things floating past. A bin, garden chairs, flip-flops, books, a football, onions, a brolly. Helicopters and rescue boats on the go all day. Tom says he sees a dead cat. I hope not.

Water everywhere downstairs now. It is over the first step on the stairs. Mum gets a call from Julie, to say the tap water could run out any minute. The Mythe water treatment plant is flooded, so they shut it down. Mum says get ready to be very smelly.

▲ *One flooded home in Tewkesbury. There were many scenes like this.*

Monday 23 July

First thing, Mr Simpson bangs on the door to say they're giving out bottled water at the Town Hall. Dad and I go to get some. We stay close to the wall until we can climb out of the water.

It's really quiet on the High Street. No traffic, and people standing around talking. One woman is crying. Most of the shops are closed. We queue at Somerfield for milk and a few more tins.

The taps are empty now. So can't wash, or flush the loo. The water company has put bowsers (water tanks) around town. You have to collect water from them in buckets. But it is not safe to drink.

Wednesday 25 July

I'm SO BORED. Stuck in all day. They won't let us out in case there's sewage in the flood water. I'm SICK OF SANDWICHES. Dad went out this evening with his buckets. Came back without any water. All the bowsers empty. He's VERY annoyed.

Friday 27 July

The water has drained away. It looks awful downstairs, and it stinks. Water got into the fridge and the kitchen units. Weeds on the carpet. I even saw worms on it. DISGUSTING. Sofa soaking. Wallpaper loose. Loads of mud and rubbish outside the front door.

A man from the insurance company arrives to check on the damage. Mum says we'll get money but not for a while.

Monday 30 July

A council truck takes away the ruined carpet and sofa and things.

Still no tap water in our street. That's a whole week! Mum says she's HAD ENOUGH, we're going to Aunt Eileen's. Dad looks extra worried and says he'll stay here until the water comes back on, and to keep burglars away.

I'm glad we're going. I'm fed up here. I can't wait to have a shower.

▲ *Guess what they're talking about?*

How insurance works

- You want to insure something.
- You decide how much it is worth.
- You ask an insurance company to insure it.
- The company says yes, and asks for a fee or **premium**. (The amount depends on what the thing is worth.)
- If the thing gets lost or damaged, the company will then pay you what it was worth, or the cost of repairs.
- They may send someone out to check before they pay.

Your turn

1 a Make a table like the one started here:

How Ellen's family got ready for the flood	
Action	Reason
Got torches	

Then fill in all the things Ellen's family did, to prepare for the flood, and say why they did it.

b Now number all the actions in your table, in order of importance. Start by putting a **1** beside the one you think is the most important.

2 Is there anything else Ellen's family could have done, to reduce the amount of flood damage in their home? See what you can come up with.

3 Imagine *your* home gets flooded with no warning, to a depth like the one shown on page 92. Go around the rooms in your mind, and list what would be ruined.

4 Now imagine you are warned that floods are on the way. You have two hours to get ready. What will you do? What will you try to save? Write a list.

7.2 What causes floods?

In this unit you will learn what causes floods – and how we humans make them worse.

What causes floods?

Most floods are caused by heavy rain. The rain quickly finds its way to the river, as this drawing shows. The river gets too full, and water flows over the banks.

And that means trouble, for people living or working near the river.

Floods can also be caused by ice or snow melting.

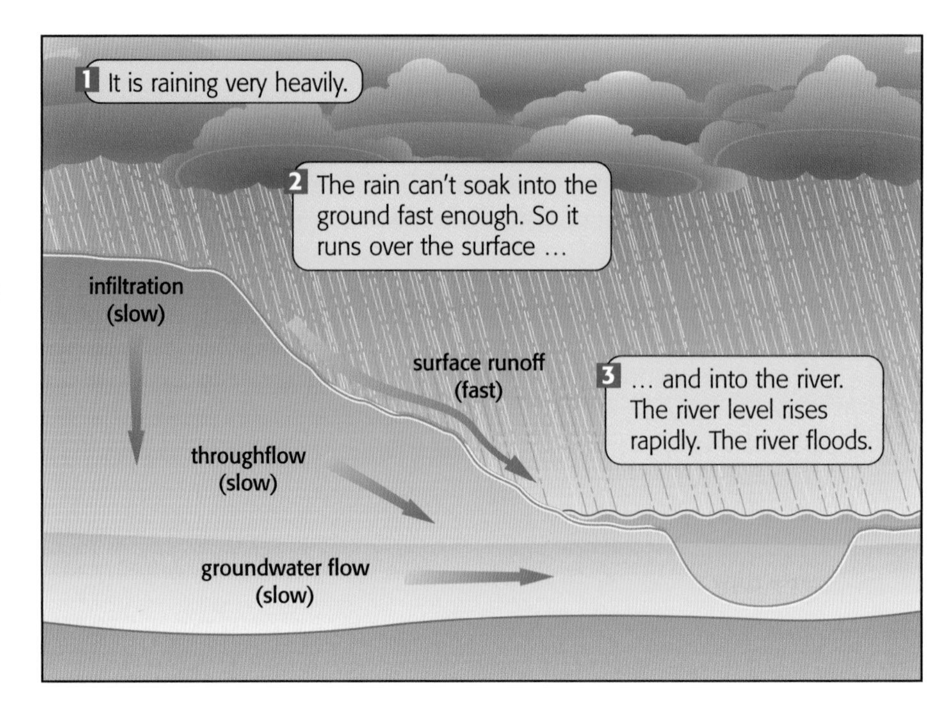

Flash floods

A burst of very heavy rain can cause a sudden flood called a **flash flood**. This happens so fast that people get no warning. They can get trapped, and drown.

Adding to the flood risk

Heavy rain is the main cause of floods. And anything that stops the rain soaking into the ground will add to the flood risk. Study the drawing on the next page. Then try these *Your turn* questions.

Your turn

1 These sentences explain how a flood occurs. They are in the wrong order. Write them in the correct order.
- The river fills up with water.
- The ground gets soaked.
- More rain runs over the ground and into the river.
- Heavy rain falls for a long period.
- The water rises over the banks.
- Infiltration slows down.

2 a Draw a spider map to show all the factors that contribute to flooding. You could start like this:

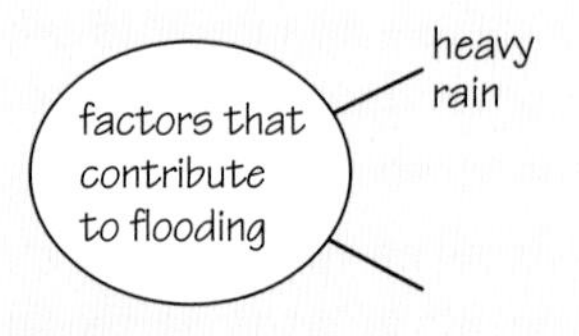

b Now underline the *natural* factors in one colour and the *human* factors in another.

c Which group of factors can we do something about?

3 Flooding is natural. All rivers flood at some time. They flood on their **flood plain**.
Pages 82 and 95 will help you with this question.

a What is a flood plain?

b A river's flood plain is … (Choose one.)
   i on high steep land near the source of the river
   ii on flat land along the river.

c In its flood plain, the river … (Choose one.)
   i usually flows very straight
   ii often has meanders.

4 a All rivers flood at some time. But it's not always a problem. Look at page 95. Why are floods:
   i not really a problem, if they occur at **X**?
   ii a big problem, when they're at **Y**?
   iii much less of a problem, if they are only at **Z**?

b Look at **Y** again. What could be done to stop floods reaching these homes? Tell us your suggestions. (You can't move people away.)

Factors that contribute to flooding

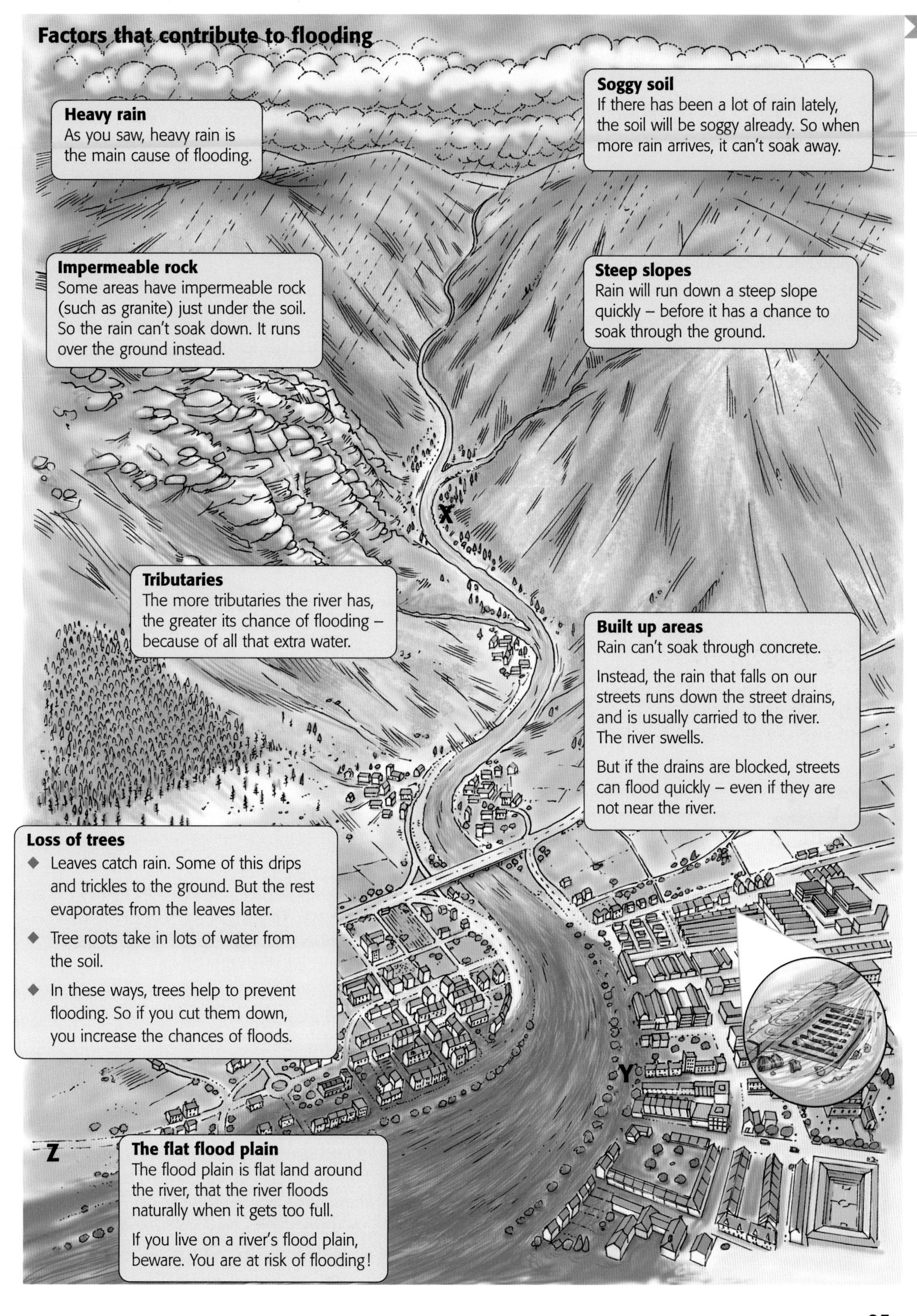

So – why did Tewkesbury flood?

In this unit you'll discover for yourself why Tewkesbury flooded so badly, in 2007.

Be a flood detective

In Unit 7.1, you read about the floods in Tewkesbury. Why did it flood so badly? Study A to F for clues. Then try *Your turn*.

Did you know?

◆ Over 2 million homes in the UK are at risk of flooding by rivers or the sea.

Wet wet wet

The months May to July of 2007 were very wet, in the UK. In fact the wettest May – July ever recorded. (We began to keep records in 1776.)

The total rainfall in those three months was nearly 40 cm, or over twice as much as usual. Imagine a layer of water nearly 40 cm deep falling on the UK!

But it did not fall equally everywhere. Some places got much more than others. Look at the map.

The heavy rain in May left the ground soggy. Then came downpours in June and July, which led to severe flooding in many areas.

Some places around Tewkesbury had up to 10 cm of rain on Friday 20th July.

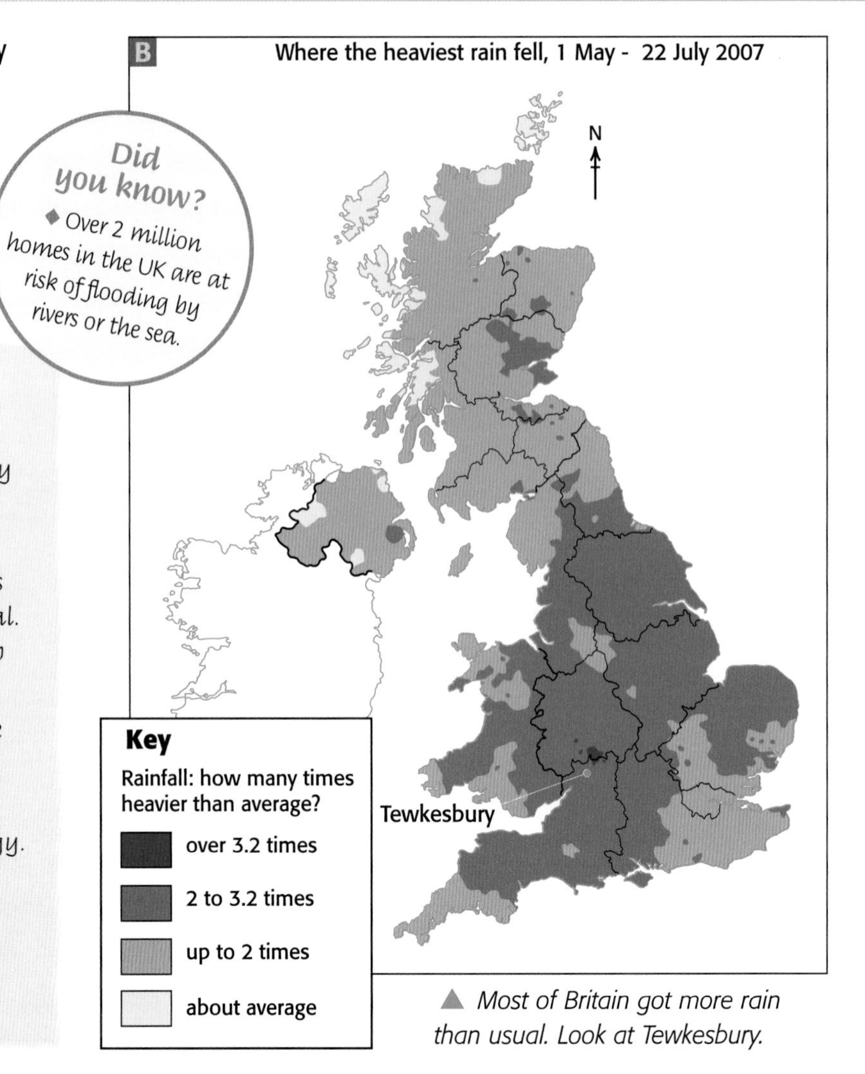

▲ *Most of Britain got more rain than usual. Look at Tewkesbury.*

Your turn

1 Tewkesbury got more rain than usual in May – July 2007. How many times more? (Check map **B**.)

2 Look at map **C**. What clue(s) can you find that Tewkesbury is at risk of flooding?

3 Say what part these played in the Tewkesbury floods. (Pages 95 and 62 may help.)
  a There was a lot of rain in the weeks before the flood.
  b The River Severn rises in the Cambrian Mountains and flows through some very mountainous land.
  c The rainfall around Tewkesbury was extra heavy on 20 July.

4 Now look at **D**, the OS map.
  a How would you describe the land at Tewkesbury?
    i mostly steep   ii mostly flat
    What is your evidence?
  b Do you think what you found in **a** played any part in the flooding? Explain your answer.

5 Look again at the OS map.
  a The Mythe waterworks is in square 8833. (It is marked as *Wks*.) It got flooded. Suggest a reason.
  b The club house (CH) on the golf course at 8831 was not flooded. Why not?

6 Now compare the OS map with the photo on page 90.
  a i First, find the Abbey on the photo. How did you identify it?
    ii Give a six-figure grid reference for it.
  b What are all those white objects in the photo, near the Abbey? (The OS key on page 138 may help.)
  c Pick out one road that got flooded, from the photo. Give a road number or grid reference for it.
  d See how many other things you can identify on the photo, using the OS map for clues.

7 Try this! In which direction was the camera facing, for:
  a the photo on page 90? b the photo on page 97?

▲ *Tewkesbury – with the floods almost gone.*

E

About Tewkesbury

Tewkesbury is a town in Gloucestershire. It stands at the confluence of two rivers: the Severn and the Avon. It has a population of about 10 000.

The 900-year-old Abbey dominates the town. There are also many fine old Tudor buildings. Their builders were wise. They built on the higher ground, to avoid floods. Many of the modern buildings are lower.

Tewkesbury grew as a market town, and a centre for milling flour. Today there is no milling, but the town attracts many tourists.

The rivers make Tewkesbury a pleasant place. But the flood risk is high – and the town has no permanent flood defences.

Did you know?

◆ *The River Severn is the UK's longest river (354 km).*

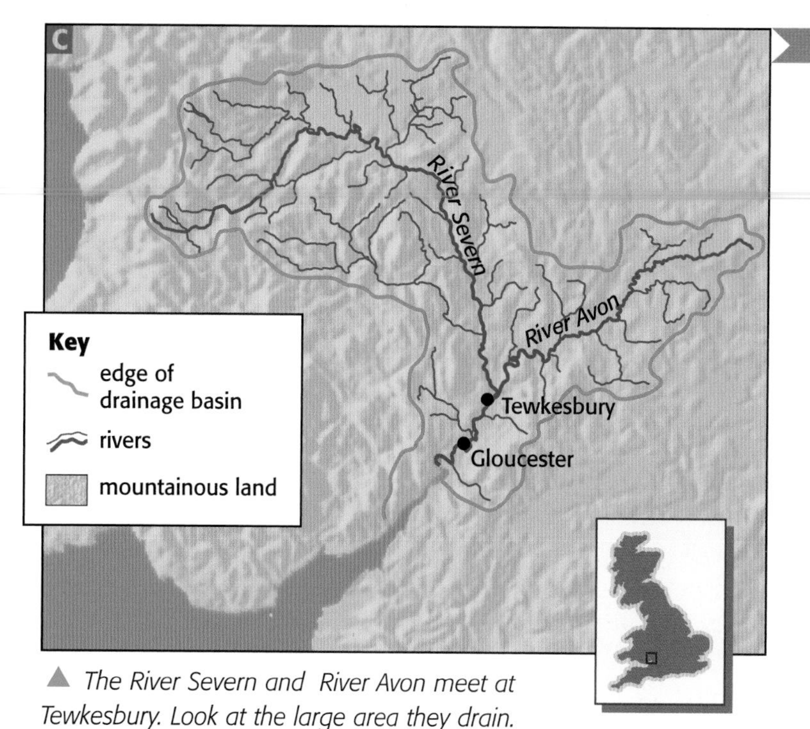

▲ *The River Severn and River Avon meet at Tewkesbury. Look at the large area they drain.*

OS map of Tewkesbury

Scale 1 cm: 500 m

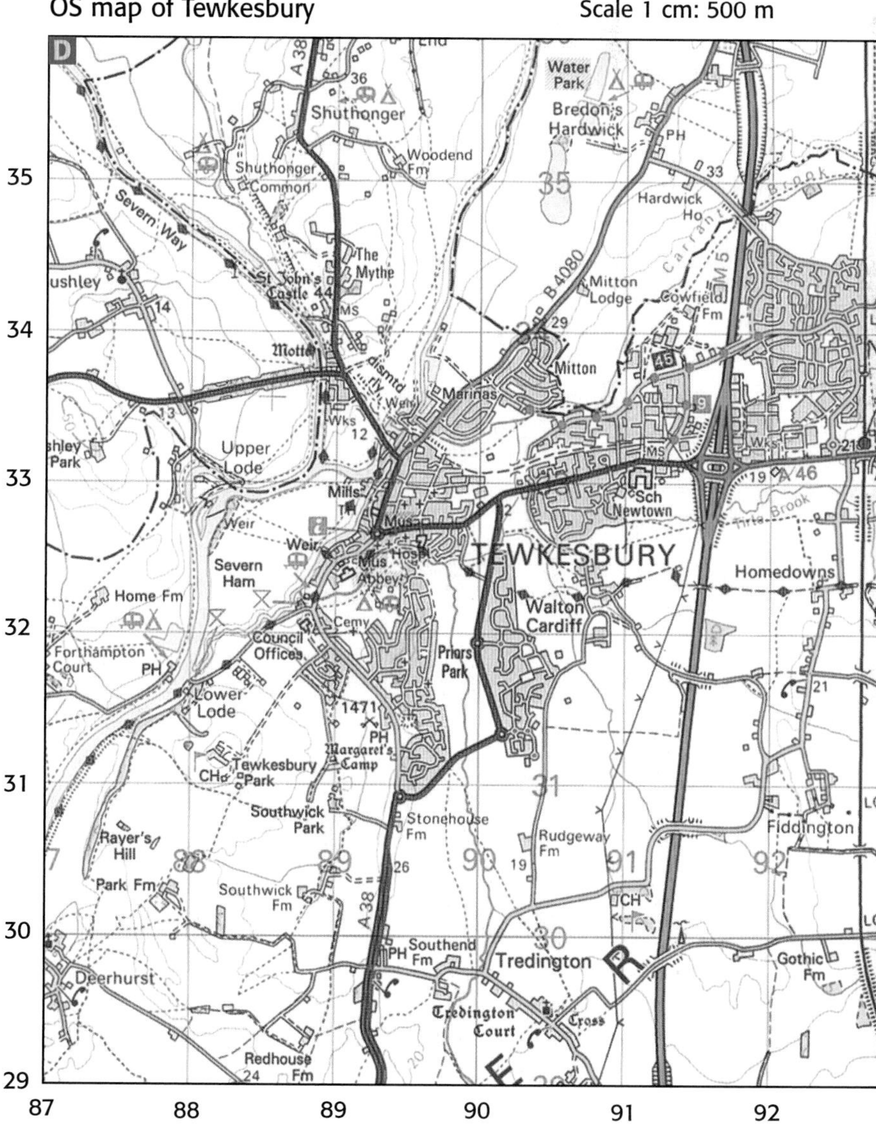

Who helps in a flooding crisis?

Here you'll find out who helps, when there is serious flooding in the UK.

You are not alone

People get help from many sources, before, during and after floods. Here are some examples from the floods in Tewkesbury.

The police They rescue flood victims trapped in homes and cars. They monitor flooded roads, and watch for looters and vandals.

The fire brigade They rescue trapped people, and animals, and pump out flooded places (like the Mythe water treatment plant).

The Environment Agency It monitors river levels, and gives flood warnings. You can ring its Floodline and check its website.

The Met Office It does weather forecasts. The Environment Agency uses these to help it decide on the flood risks.

The army It is often called in, in an emergency. In Tewkesbury, soldiers helped to deliver bottled water around town.

The RAF and Coastguard Agency They provide helicopters and teams to rescue people in danger. They were kept busy in Tewkesbury.

Gold Command It's the police HQ that takes overall charge of rescue services, in an emergency. (For this area, it's in Gloucester.)

The local council Its workers give out sandbags when there's a flood warning. After floods they clear the debris and repair roads and bridges.

Local schools, churches, clubs, and the Red Cross They provide food and shelter for flood victims. The Red Cross also raises money.

The government It stays in close touch with Gold Command. It may give money to help the flooded area, and flood victims.

Local shops and other businesses They give free food, drinks, and other things flood victims need – or money for these.

Insurance companies They send out assessors to check the flood damage in insured buildings. Later they pay compensation.

Your turn

1 Look at all the arrows above. What do you think they are telling you?

2 From the groups of people that help, name:
  a three that provide help even before floods arrive. How do they do this?
  b two that may still be doing work related to the floods, weeks after the floods have gone
  c two groups that help for free
  d three that are paid for their time
  e three that are *emergency services* (Glossary?)
  f two that belong to the *defence forces*
  g one that's a *non-governmental organisation (NGO)*
  h one that helps only the people who had paid them

3 Which group above do you think has the *most* important role, in times of flooding? Explain your choice.

4 You are Ellen. Your teacher has asked you to interview one of the groups above, about the Tewkesbury floods. Choose one, and write down the questions you'll ask.

7.5 Flooding: the consequences

Floods can affect all of us – not just the people who live in the flooded areas. Find out how, in this unit.

The flood damage in the UK, 2007

Tewkesbury was not the only place to suffer, in the wet summer of 2007. Many places were flooded. This map shows some of them.

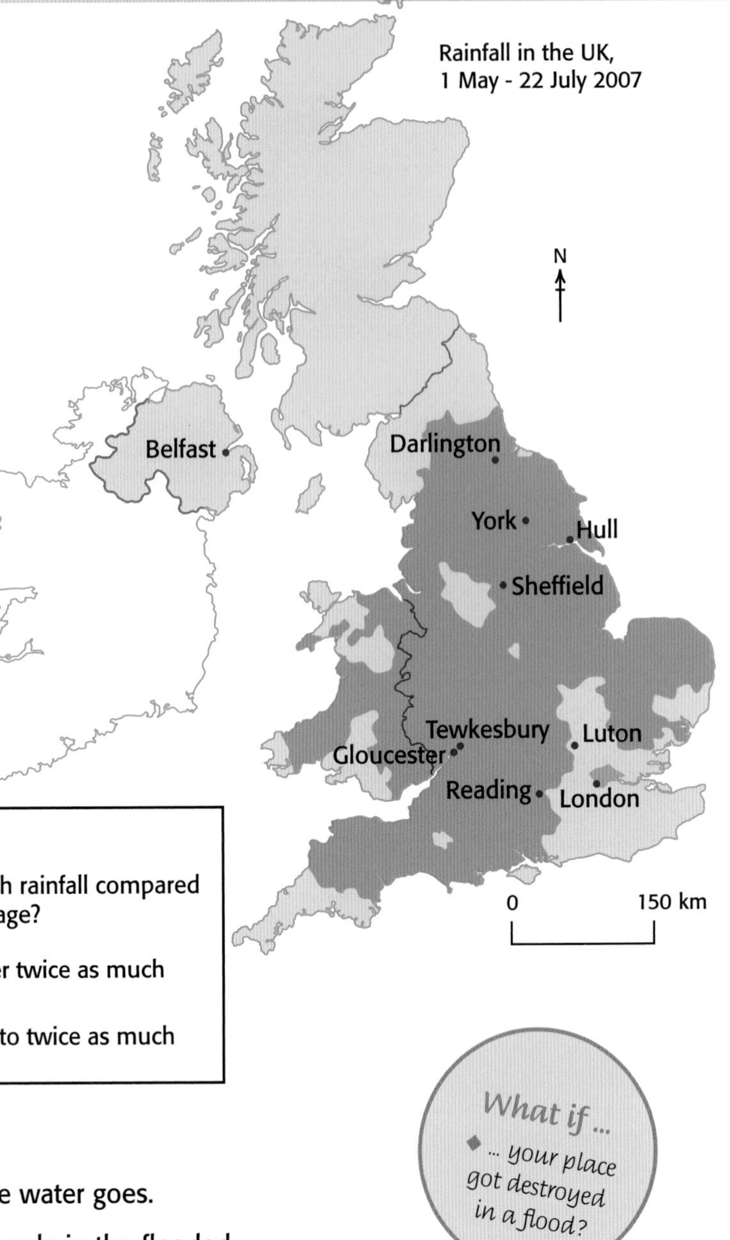

The flood damage around the UK

- 10 people died, by drowning or in flood-related incidents.
- Many thousands of homes and businesses were damaged by flood water.
- Roads, railway lines and bridges needed repair.
- Many farms lost animals and crops to the floods.
- The cost of the damage was around £6 billion.

What if ...

- *... we had this much damage every year from floods?*

What if ...

- *... your place got destroyed in a flood?*

The consequences

The problems caused by flooding don't stop when the water goes.

The 2007 floods had serious consequences for the people in the flooded areas. And also for everyone else in the UK. Look at the next page for some examples.

Could they happen again?

Scientists predict worse storms and floods in the UK in the future, thanks to global warming.

That's grim news. We need to be prepared. In the next unit you can read about ways to protect ourselves from floods.

The trouble is, lots of our cities, towns, and villages, and much of our **infrastructure** – railways, roads, power stations, water treatment plants, sewage works – are in flood plains. Protecting them will cost £ billions.

If the government has to spend all that money fighting floods, it could mean less for schools, or hospitals, or fighting crime. Or people will have to pay more taxes. Hard choices will have to be made.

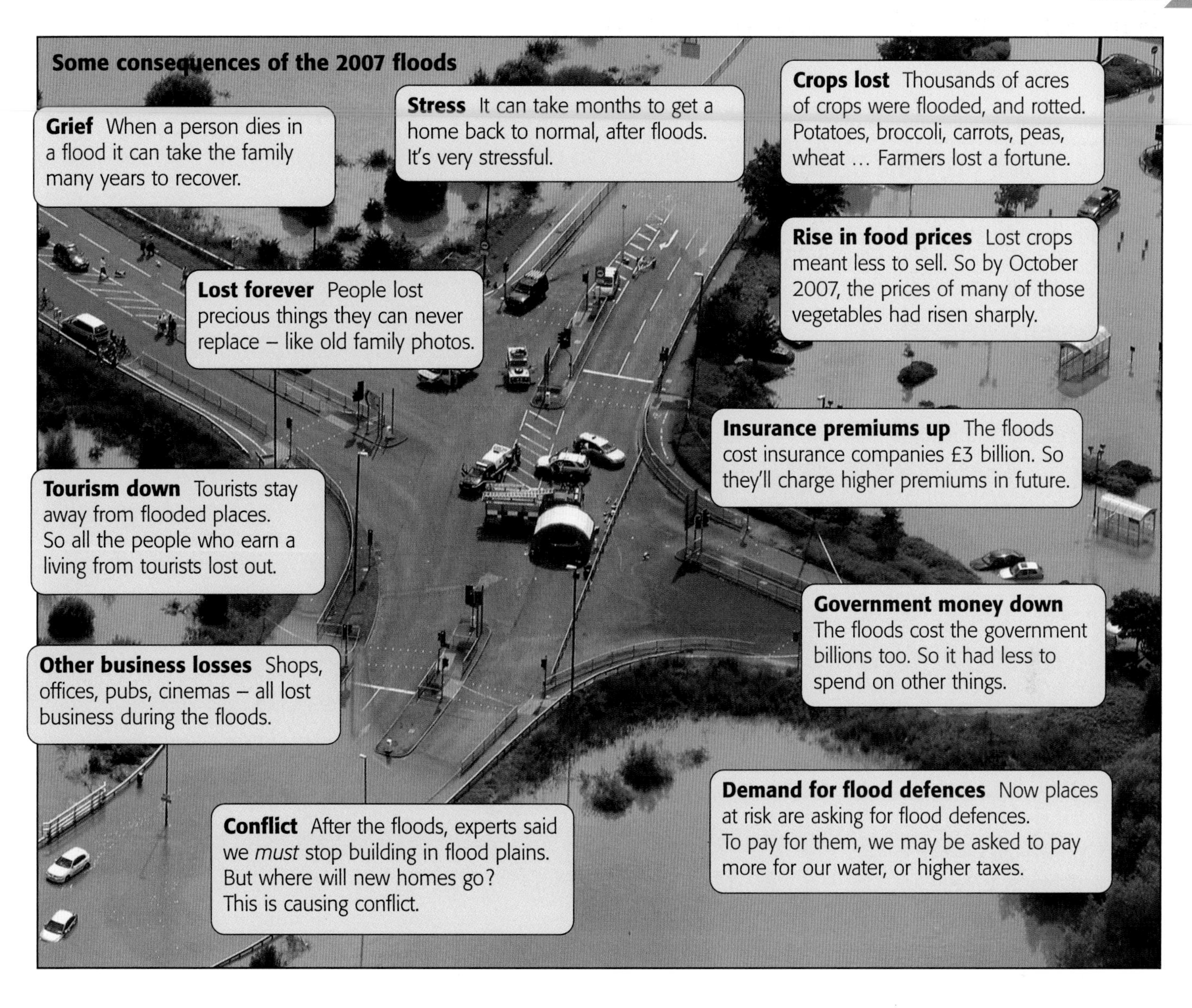

Your turn

1 Look at the flooded places named on the map on page 100. They are all on flood plains. *Why* are so many of our settlements on flood plains? (Page 36 may help.)

2 Look at the consequences of flooding, above. See if you can pick out:
  a two that affect just flood victims and their families
  b two that affect all of a flooded area
  c two that affect the whole UK
  d two emotional consequences
  e four financial consequences
  f a social consequence
  The glossary may help.

3 Some consequences of flooding are **short-term**. They could affect us for say 6 months or less. Others are **long-term**. They can last for years.
  a See if you can divide the consequences above into two lists, like this:

Short-term	Long-term
rise in food prices	

  b Did you find it easy to do? Explain.

4 Look at the long-term consequences of the 2007 floods in your list for 3a. Underline any that may affect *you* when you are an adult.

7.6 Protecting ourselves from floods

In this unit, you'll learn about ways to prevent floods.

How can we prevent floods?

When rivers flood, we help the victims with shelter and food. But that's a short-term solution. We also try to stop the floods happening again. Here are four ways to do that:

1 Control the water level

- Build a **dam** on the river to trap water. You can let the water out slowly. (Dams can be large or small.)
- Build **pumping stations**. Then when the water level rises, you can pump water out of the river and into storage basins, or even onto empty fields.

2 Build flood defences

- Build up the river banks to make **embankments**, to keep water in.
- Or build **flood barriers** around built-up areas to keep water out. (These could be concrete walls, or metal barriers you put up for floods and take down later.)

3 Make the river channel bigger

You could dredge material from the river bed and banks, so that the channel will hold more water.

4 Improve street drainage

- Make sure street drains can cope with heavy rainfall.
- Make sure they are cleaned often, to remove any blockages.
- If the water from them could be drained into soil, instead of the river, the river would not rise so fast.

5 Control land use around the river

- Stop people building on the flood plain.
- Plant more trees in the drainage basin.
- Pay farmers to allow fields along the river to get flooded. (So there will be less flooding elsewhere.)

Overall, preventing floods can be a very expensive business.

Protecting our own homes

Lots of us live on flood plains already. And people are still building new homes on them, in spite of the flood risk.

So what can we do to homes, to protect them? This drawing gives some ideas. What do you think?

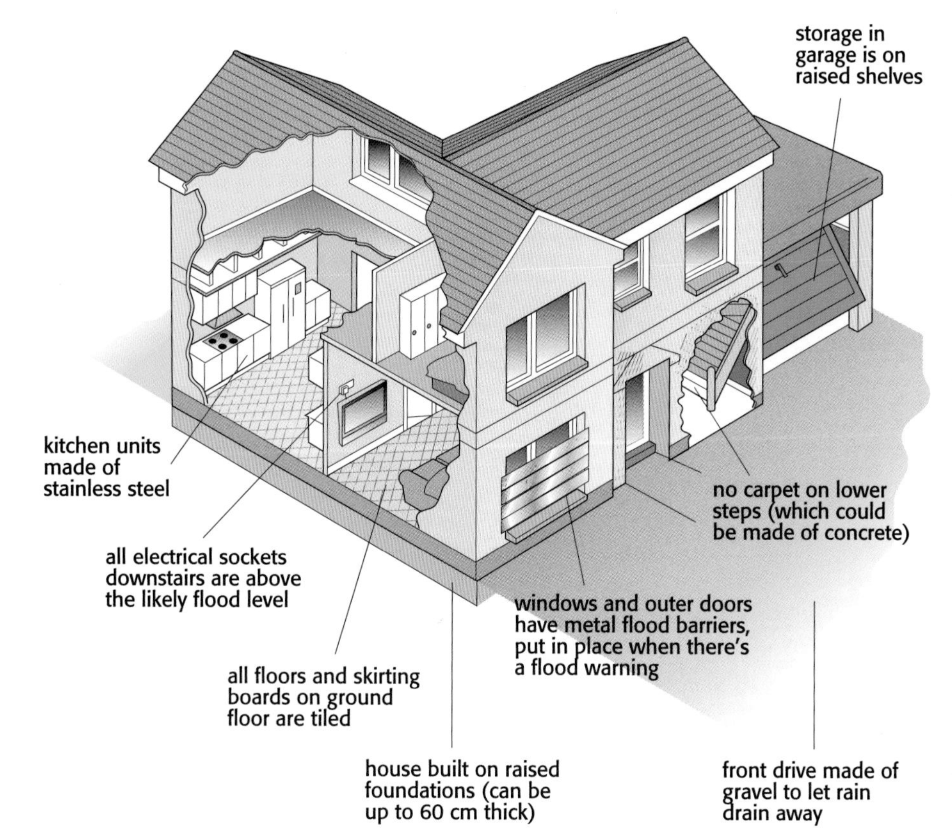

▲ *Temporary flood barriers in place.*

▲ *An embankment along the river.*

Your turn

1 Look at the drawing on page 102.
  a Name the features at **A**, **B**, **C** and **D**, and say what job each does.
  b What is going on at **E**? How can this prevent floods?

2 Each blue panel on page 102 shows one approach to flood prevention. Which approach do you think:
  a costs most? b is best for wildlife?
  c costs least? d makes the river look best?
  e would be best if you were building a new town?
  Give reasons for your answers.

3 Tewkesbury has no permanent flood defences. On the right are proposals for protecting it in future. They use OS grid references.
  Using the OS map on page 97 and the photos on pages 90 and 97 to help you, give each a score from 0 to 5. (0 = very bad idea; 5 = great idea.)
  Then give your reasons for each score.

4 Tewkesbury Council wants ideas for protecting the Mythe water treatment plant (see 8833) from flooding. (You can't move it.) What do you suggest?

5 Look at the drawing of the flood-proofed house, above.
  a Pick out what you think are the two best ideas on it, and say why you chose them.
  b Now see if you can come up with other ideas. (For example, houses that float? A place on the roof where you wait for a helicopter to rescue you?)

A *Plant a forest in square 8732.*
B *Build a dam across the River Severn in 8735.*
C *Build a tall flood wall between the buildings and the water, from 8832 up to 8933.*
D *Have temporary flood barriers ready, to go from 8832 to 8933. Store them at the council offices in lower 8832.*
E *Build two pumping stations with concrete storage basins, one in 8834 and one in 8934.*
F *Remove all buildings from the flood plain at Tewkesbury.*

bet
24
SAMSUNG
mobile

The big picture

Geography is brilliant – it even covers sport! These are the big ideas behind this chapter:

- Sport links people and places all around the world.
- Some sports depend on the natural environment – and some on the venues we build.
- Sports are big business. Some sports people earn a fortune.
- But the people in poorer countries, who make sports kit, get paid very little.
- Sports can have a big impact on an area, for better or worse.

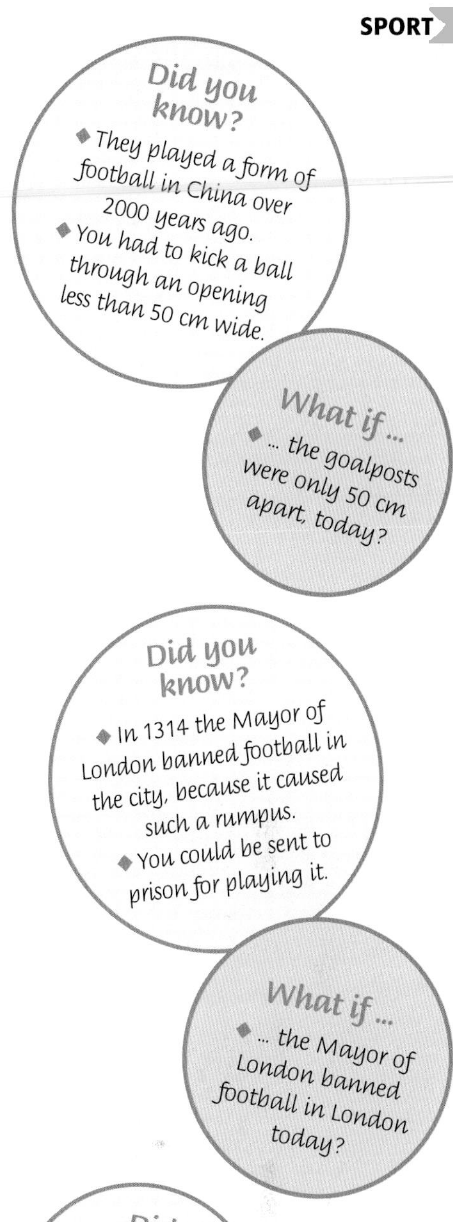

Your goals for this chapter

By the end of this chapter you should be able to answer these questions:

- In what ways does a sport (such as football) link people and places around the world?
- What examples can I give of sports that depend on:
  the natural environment? built venues?
  (At least three for each.)
- Football is a business, as well as a sport. Why?
- What factors would I think about, in choosing a site for a new sports stadium? (At least five.)
- Why is most sports kit made in poorer countries?
- Where in London will the 2012 Olympic and Paralympic Games be held? What was the site like before? And what will happen to it after the Games?
- The planners aim to make the 2012 Olympics sustainable. What does that mean? And how will they do it? Give at least six examples.

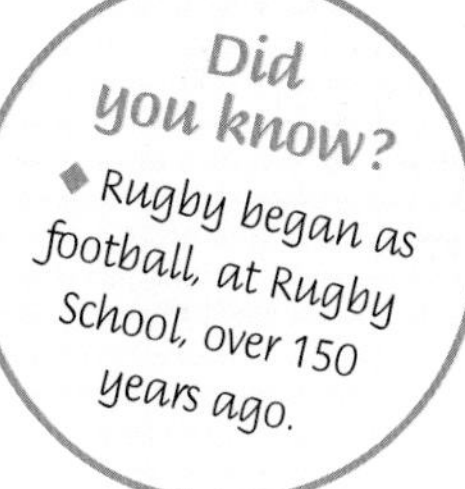

And then ...

When you finish the chapter, come back to this page and see if you have met your goals!

Your chapter starter

Football is just geography in action!

Look at the photo on page 104.

The ball was made in Pakistan, and the boots in India. The match may be watched on TV in Nigeria, and Japan, and other countries.

What other links to geography can you find in the photo? See how many you can come up with.

Geography and sport

In this unit you'll explore the links between geography and sport.

What's the connection?

What has geography got to do with sport? You're about to find out! Look at these photos. Then try *Your turn*.

A

B

C

D

E

Your turn

These questions are about the sports in the photos.

1 a First, name the sports shown in photos A – I. (Easy!)
b Which one would you most like to take part in? Why?
c Which would you least like to take part in? Why?

2 Which of them usually depend on natural features, and the natural environment? (For example, on the wind…) For each one you list, explain why.

3 a For which of the sports do we build special venues?
b Which of these venues are usually in towns or cities?

4 For each photo, say how a cold, wet, windy, day would affect the people, or the event.

5 a Now for each sport, try to name one place where you could do it, in the UK.
b And, as an extra challenge:
i draw a rough sketch map of the UK
ii see if you can mark your places from **a** on it!

6 For each sport, see if you can name a special event, where people or teams from different countries compete. Or a special challenge people have in mind. (An example for football would be the World Cup.)

7 So… is there a link between geography and sport? Decide *Yes* or *No*. Give your reasons as bullet points.

The football business

Sport is fun – and big big business. Here you will explore it as a business, with football as our example.

Did you know?

- Football's top transfer fee so far is £45.62 million ...
- ... paid in 2001 by Real Madrid for Zinedine Zidane.

It's big business

Football is not just a game – it is big business. The top clubs earn, and spend, millions of pounds a year. These photos give clues about how they earn money. (You will match the numbers to words later.)

How clubs spend money

- paying the players
- paying other staff
- buying new players
- improving stadiums
- hosting matches
- going to away matches
- training youth teams
- working with schools

Now look at the list on the right. It shows how clubs spend money. Some get into trouble because they spend more than they earn!

And not just for the clubs

It is not just football clubs that earn money from football.

Local shops, cafes, pubs and restaurants are all busier than usual on match days.

They'll be ever so hungry and thirsty after this.

Your turn

How a big football club earns money

	Ways to earn money	Could the club earn more from this by … having better players?	moving to a bigger, better stadium?
1	sell match tickets		yes
2			

1 This box gives ways a big football club earns money: This is what you have to do.

- sell match tickets
- catering (bars and restaurants)
- sell merchandise (strip, scarves, and so on)
- TV fees (for matches shown on TV)
- rent out rooms for conferences
- rent out private viewing boxes
- sponsorship

a Make a table with headings like the one started above. Give rows numbered 1 to 7, to match the numbers on the photos on page 108.
b In column 2, fill in a way to earn money, to match what's in that photo. (Pick from the box above.)
c Look at row 1. Will a club sell more tickets, if it has better players? Write *yes* or *no* in column 3.
d Now fill in column 3 for the other rows.
e Think about each way to earn money, in turn. Will the club earn more from it, if it moves to a better stadium? Decide, and fill in column 4.

2 Who else gains when a club is successful? Show them on a spider map. You could start like this:

3 Like every business, football clubs need to earn money. Many buy players from all over the world, to help them. This flowchart will show how it works:

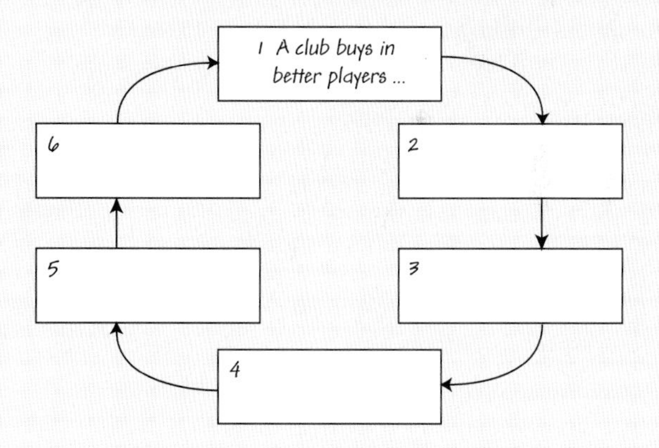

Make a larger copy of the flowchart. Then write these in the correct boxes …

… so it can afford more of the world's top players.
… which means it sells more tickets and more merchandise.
So the club gets richer and richer…
… so it wins more and more matches…
This also means it makes more money from TV and sponsorship.

4 Now back to stadiums. Around the UK, many clubs have moved to new stadiums, or plan to. Using your table for 1 to help you, write a paragraph to explain why.

Liverpool FC is moving home

Here you can explore Liverpool FC's plans for a new stadium.

A new stadium for Liverpool FC

'I'm blown away' said Steven Gerrard.
'Spectacular' said Jamie Carragher.
'I'm really looking forward to playing there' said Xabi Alonso.
These are their reactions to the plans for Liverpool FC's new stadium, which were shown to the players today.

Since 1892, when the club started, Anfield has been its home. But from 2010, home will be a stunning new stadium in Stanley Park, just a short walk away. The new stadium will cost over £400 million. The old one will be redeveloped.

'This is a big step in our plans to regenerate the Anfield area' said a Liverpool City Council official. 'It's quite run down. But not for much longer. There are exciting changes ahead.'

Everyone agrees the club needs a new stadium. But not everyone is happy about the new location. There have been over 400 objections so far.

Based on news reports, August 2007

Letter to the editor

Dear Sir

Stanley Park is a historic old park – and a welcome green space in this built-up area.

We must not let Liverpool FC build here. A glitzy new stadium, nearly 80 m high, will be so out of place.

The council promised not to sell off our parks. Now it is leasing parkland to a football club, for 999 years. Is that any better?

Liverpool has lots of brownfield sites. The club should go find one, and leave our park alone.

Yours

Len Williams

Your turn

1 What do you think the club's reasons are, for moving? (Your work in the last unit may help you answer.)

2 When Liverpool FC first thought of moving, these two fans came up with proposals:

Suggest reasons why the club said no to each one. The photo on page 111 may help for **A**.

3 a Make a table like this:

Finding a new site for a football stadium	
Things to think about	**Score for Stanley Park**
Large enough?	
Close to public transport?	
Near the old one, for our fans' sake?	

b In the first column, list more things you'd think about, if you had to find a new site for a stadium. What about cost? The environment? Anything else?

c In the second column, give Stanley Park a score for each. (0 = poor, 5 = excellent.)

4 It's your job to decide *where* in Stanley Park the new stadium should go. (Let's assume the stadium will measure 200 m by 150 m.)

a Make a sketch of the park. The OS map will help.

b Mark in where you will put the new stadium.

c Mark in where you will put the entrance road(s) to it.

d Add notes giving the reasons for your decisions.

5 Next, you have to redevelop the old stadium.

a Turn to page 50, and remind yourself what *sustainable development* means.

b Now think up some sustainable plans for the old stadium. Show your ideas as simple labelled sketches. These facts may help:

- There is not much work locally, for people.
- There is a housing shortage in the UK.
- There's a shortage of leisure facilities in the area.
- The old stadium has a long and rich football history.
- Fans really love the old stadium.

6 Finally, you live at **X** on the OS map. You are not mad about football. You like to walk your dog in the park.

a See if you can find your house on the aerial photo.

b How do *you* feel about the club's move? Say why.

c Write your own letter to the paper about it.

The proposed new stadium

- It will hold 60 000 people.
- It will cost over £400 million.
- The club will pay Liverpool City Council a rent of £300 000 a year for the land.
- The council will use the money to help Liverpool.

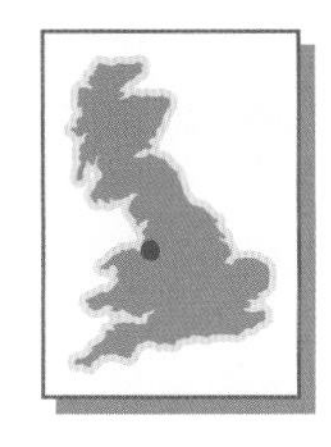

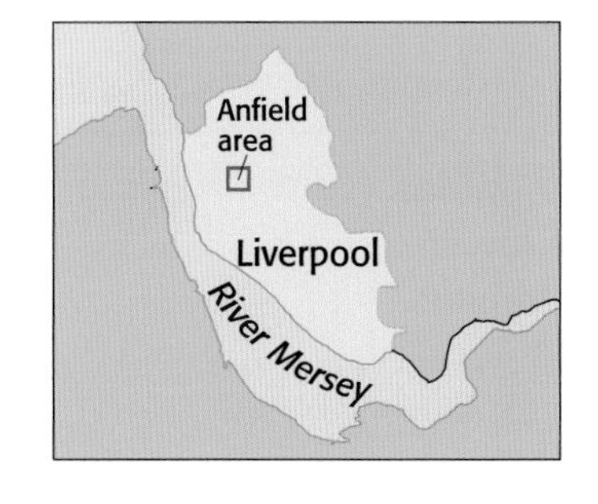

The Anfield area in Liverpool, in 2006

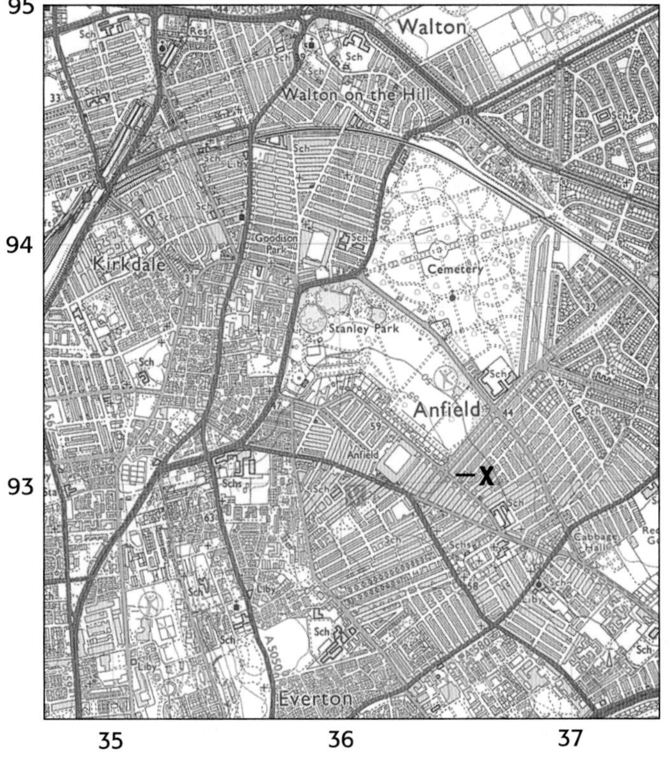

Scale 1 cm : 250 m

Who are the losers?

Sport is big business – but not everyone is a winner. Here you'll find out about the people who make sports goods. We take footballs as our example.

Football skills

This player earned around £1500 today for kicking this ball. He scored a goal, with great skill.

The football is made of rubber and synthetic leather. It is top quality. It could cost £65 in the shops.

It was sewn by hand by Omar, with great skill. It took him 3 hours. He got paid 65p. He is aged 14.

Omar lives in a village near Sialkot, in Pakistan.

In that city, and the villages around it, they know everything about footballs. They make over 80% of the world's hand-stitched footballs. They sew over 40 million footballs a year!

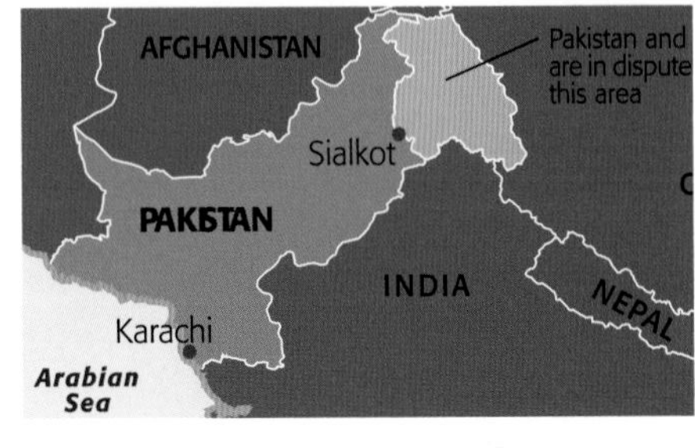

A vicious circle

Why is Omar paid so little for his skill? This is what happens:

This British company supplies footballs. It gets them made ...

... and then sells them to clubs and sports shops at a profit.

The less it pays for the footballs the more profit it will make ...

... so it searches the world for a factory to make them cheaply.

This factory owner in Sialkot wants the work. If he charges too much he won't get it!

He also wants to make as much profit as he can. So he pays his workers very little.

They can't leave because they need the work. If they don't have jobs they will starve.

The footballs arrive in the UK. They have been well made. The British company is happy.

It's not just British companies. It happens in all the richer countries. They get things made in poorer countries where wages are lower. Not just footballs, but football boots, and cricket bats, and trainers, and sports clothing. There will be more about this in geog.3.

Omar's story

I have been sewing footballs since I was 8.

I don't like it much. But I have to do it, because my dad died and we need the money. My mum used to sew too but now she has trouble with her eyes. So I have to support my family.

I work in a stitching centre. I start at 7 in the morning and often work till 8 in the evening. I do 4 footballs a day – so I earn £2.60 a day. But they can throw me out any time. I don't know what I will do if that happens.

I get tired of sewing all day. My shoulders get stiff. My eyes get sore. My fingers are all cut. I would love to go to school instead – but no chance!

I saw a World Cup match on the TV at my uncle's house. The football could have been one I sewed. But nobody at the match knew about me!

▲ *Omar at work.*

Your turn

1 A certain Premier League footballer earns £15 000 a week. (The really big stars earn a lot more.) How long would Omar have to work to earn that much?

2 Suppose you buy a football for £50, made in Sialkot. Where does the money go? It could go like this:

	£
The shop where you bought it	10.00
The British supplier	31.00
Shipping and transport companies	1.50
The Sialkot factory owner	5.00
The company that supplied the materials for the football	1.50
The stitcher	0.50
Other factory costs (lighting etc)	0.50
Total	**£50.00**

What percentage of the money does the factory owner get? You can work it out like this:

$$\frac{\text{factory owner's share}}{\text{total amount}} \times 100\%$$

$$= \frac{£5}{£50} \times 100 = 10\%$$

a Now work out the percentages for the others in the list.

b Draw a pie chart to show how the money is shared.

c Who gets: the largest share? the smallest?

3 Look at the 'vicious circle' on page 112. What might happen if:

a the stitchers went on strike?

b the factory manager tried to charge the British company more?

c the British company charged the shops more?

d everyone refused to use footballs made in Pakistan?

e someone invented a machine that could sew footballs perfectly?

4 In some ways, Omar's life is also a vicious circle.

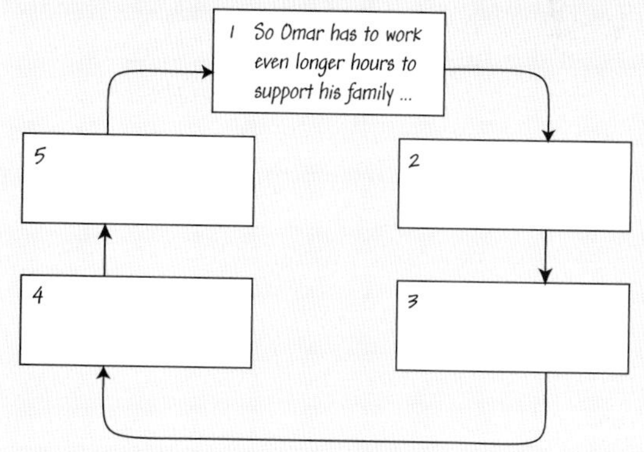

Make a larger copy of the drawing above, Then write the following in the correct boxes.

... but every year food and clothes cost a little more ...

... so he can't get a better-paid job ...

... so he has even less chance to go to school ...

... so he can't learn new skills (like reading and writing) ...

5 How would you make life less unfair for the stitchers? Give your answer in not less than 150 words.

8.5 London 2012

Sport can have a huge impact on a place – and especially the Olympic Games. Here you can find out about London's new Olympic Park for 2012.

London's big win

On 6 July 2005, London heard some exciting news. It had won its bid to host the 2012 Olympic and Paralympic Games.

This is not just a chance to host these famous events. It is also a chance to **regenerate** a run-down area of London, to benefit people for years to come.

That area is the Lower Lea Valley in the east of London. It will be home to the Olympic Park.

The Olympic Park

This shows the Lower Lea Valley in 2005. The site chosen for the Park once had gas works, chemical industries, and a huge rubbish dump. Most were gone, but the soil was badly **contaminated**.

Over 200 businesses, like this one, were left. And a run-down housing estate, and two travellers' camps. But by 2007, the people had accepted other places, and the land was bought for the Olympics.

This is how the Olympic Park will look, in 2012. With world-class sports venues. Flats for the athletes and officials. Places to eat in and shop in. Big outdoor screens, where you can watch all the action.

And here's the big prize. When the Games are over, people will buy the flats, and use the sports venues. Businesses will move in. And a once-neglected area will be alive again.

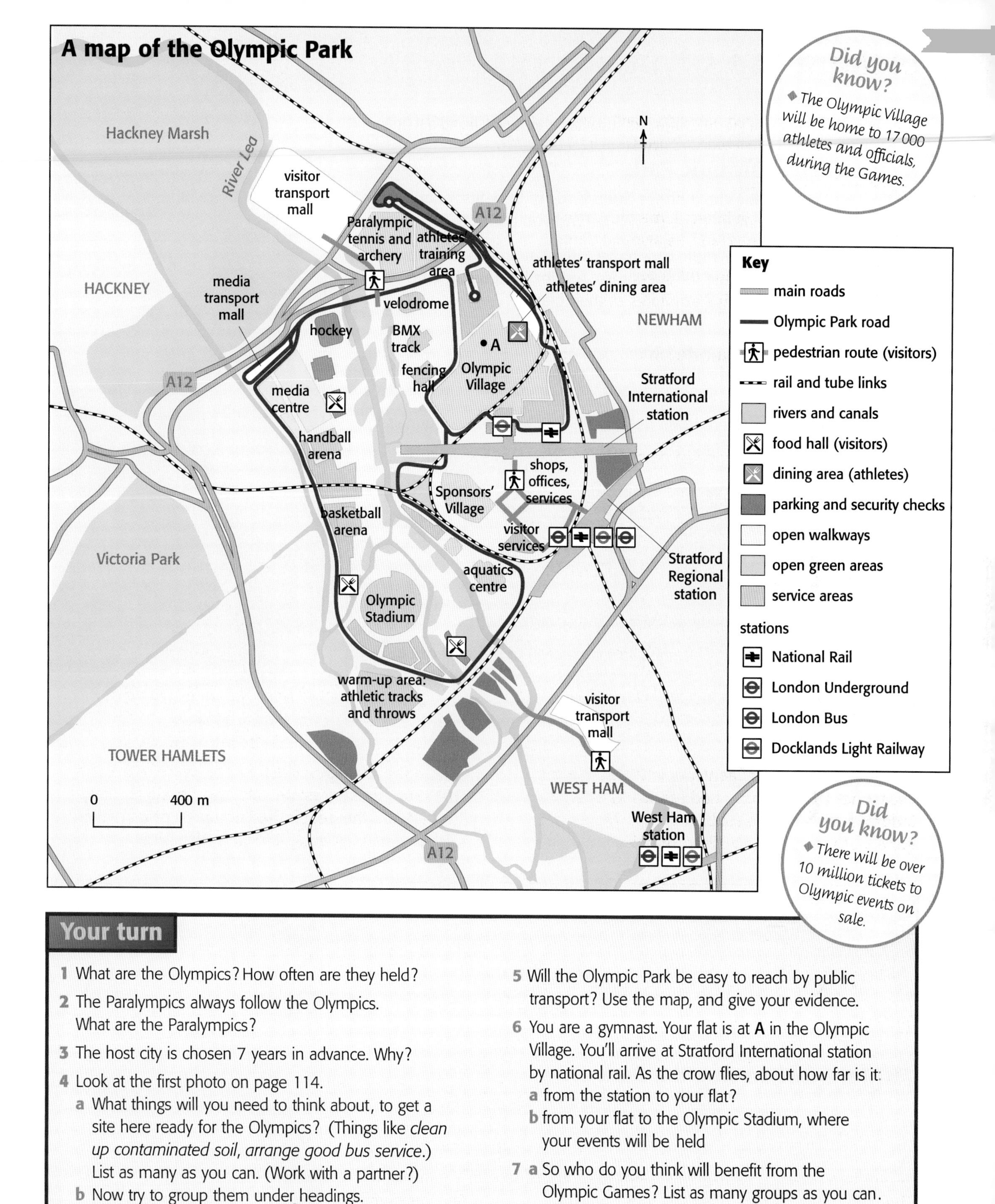

Your turn

1 What are the Olympics? How often are they held?

2 The Paralympics always follow the Olympics. What are the Paralympics?

3 The host city is chosen 7 years in advance. Why?

4 Look at the first photo on page 114.
  a What things will you need to think about, to get a site here ready for the Olympics? (Things like *clean up contaminated soil, arrange good bus service.*) List as many as you can. (Work with a partner?)
  b Now try to group them under headings. (*Clearing site* could be one heading. *Transport* could be another.)

5 Will the Olympic Park be easy to reach by public transport? Use the map, and give your evidence.

6 You are a gymnast. Your flat is at **A** in the Olympic Village. You'll arrive at Stratford International station by national rail. As the crow flies, about how far is it:
  a from the station to your flat?
  b from your flat to the Olympic Stadium, where your events will be held

7 a So who do you think will benefit from the Olympic Games? List as many groups as you can. (Will athletes benefit? Will plumbers? Will you?)
  b Who might feel they have lost out?

Making the Olympics sustainable

The planners aim to make the 2012 Olympics sustainable. Here you'll find out how.

What if ...
- ... London had lost the bid to host the Olympics?

A sustainable Olympic Park

If something is **sustainable**, it brings three kinds of benefits: **economic**, **social** and **environmental**. Page 50 will remind you about these.

The planners want the Olympic Park to be truly sustainable. So they are planning carefully – not just for the games, but for the people who will move in later. Look at these examples of their plans.

A Getting the site ready

1. The contaminated soil dug up, washed, and reused.
2. All the old pylons taken down. (New cables will be buried in tunnels.)
3. All the old buildings knocked down. Rubble reused on site later.
4. Polluted rivers and waterways cleaned up.

B The infrastructure

1. New roads, and electricity, gas, water, and sewage systems.
2. A power plant that uses **biomass** (plant material) to make electricity.
3. A wind turbine, for extra electricity.

C Work on the buildings

1. World-class sports venues.
2. Some venues will be easy to take down again.
3. Venues easy for athletes to get to, from the Olympic Village.
4. Venues cater for visitors of all ages, and people with disabilities.
5. Buildings will waste as little energy as possible. (Heat, light etc.)
6. Up to 9000 building workers employed.

D Transport serving the Park

1. Existing rail and Underground lines improved.
2. A new link to Europe, via Eurostar.
3. A new fast rail service: from the Park to central London in 7 minutes.
4. The Docklands Light Railway (trains with no drivers) extends to Stratford.
5. There is good access by road to the Park.

▲ *Down it comes, to make way for the Olympic Stadium.*

▲ *New trains like this will carry you from the Park to central London in minutes.*

E During the Games

1. Tickets won't cost too much.
2. Visitors arrive by public transport, bike or on foot. (To cut pollution.)
3. All waste from shops, restaurants and cafes will be recycled. None will go to rubbish dumps. (Waste food will make compost.)
4. Visitors really enjoy themselves!

▲ *There's basketball in the Olympics ...*

F After the Games

1. Some venues will be moved to more suitable sites in the UK. For example, the water polo venue.
2. Others, like the swimming pools, and BMX track, open for us all to use.
3. Housing sold or rented – some to people on low wages.
4. Businesses move into some buildings, and provide jobs for people.
5. More homes, shops, and offices to be built in the Park.

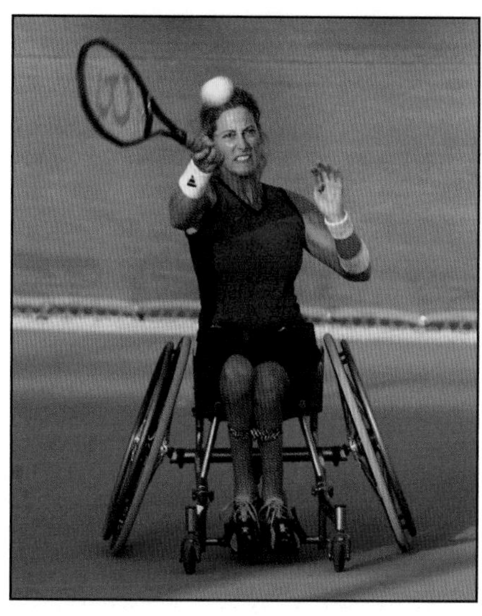

▲ *... and tennis in the Paralympics.*

And the cost?

Hosting the 2012 Olympics will cost over £9 billion!
The government, and National Lottery, will fund most of this.
And people in London are paying extra council tax (38p a week).

Money will come in again from selling tickets and merchandise.
And from catering, sponsorship, and TV rights. And later, from selling and renting the buildings as offices and homes.

London hopes the Games will be great for the UK's image too – and bring lots more tourists here in the future.

And it hopes that the lasting legacy of the Games will be ... a vibrant Olympic Park.

Your turn

1 First, turn to page 50 to remind yourself what *sustainable development* means. Then explain it to us.

2 Look at all the plans. Say which one(s) might benefit:
- a Ray, a disabled athlete competing in the Games
- b Sam, a local electrician, looking for work in 2010
- c Ita, a basketball fan, living in Belgium
- d Jenny, who lives near the Park and does not like sport, but works in central London
- e Will, a sports-mad student from Bristol
- f local wildlife
- g Abuya, who lives in Kenya, has no interest in the Olympics, and will never visit the UK

3 Some benefits are *short-term*. They last for days or weeks. Others are *long-term*. They last for years.
- a Make a large copy of this table.
- b Then see if you can fill in one example for each kind of benefit, for the Olympics.

Benefit	Short term	Long term
Economic		
Social		
Environmental		

4 Look at this person's opinion. What will you say in reply?

2012 Olympics? A huge waste of money.

9 Our restless planet

The big picture

This chapter is all about earthquakes and volcanoes. These are the big ideas behind the chapter:

- Earthquakes and volcanic eruptions have killed millions of people, and ruined millions of lives.
- They are caused by currents of hot soft rock inside the Earth, which drag big slabs of the Earth's crust around. (We call these slabs **plates**.)
- We can't prevent them. All we can do is help the survivors, and find ways to protect people in the future.
- This can cost a lot of money. But poor countries don't have much, so their people may suffer more harm.

Your goals for this chapter

By the end of this chapter you should be able to answer these questions:

- What do these terms mean ?

  crust mantle core lithosphere convection current oceanic crust continental crust
- What are the Earth's plates, and why do they move ?
- What causes earthquakes, and what kind of damage do they do ?
- What do these terms mean ?

  fault focus epicentre seismic wave aftershock tsunami
- What causes tsunami? And what kind of damage do they do?
- What are volcanoes, and what kind of damage do eruptions do?
- What do these terms mean ?

  magma lava crater pyroclastic flow mudflow ash
- What's the link between plates, earthquakes, and volcanoes?
- How do humans respond to earthquakes and volcanic eruptions?
- Why might these events cause more deaths in poor countries?

And then …

When you finish the chapter, come back to this page and see if you have met your goals !

Did you know?

- As you sit there, on your chair, you are moving very slowly eastwards – at about 1 cm a year!

Did you know?

- In 1556, an earthquake in China killed 830 000 people.

Did you know?

- In 1943, a volcano appeared in a field, in Paricutin in Mexico.
- It erupted non-stop for nine years.

What if …

- … a volcano appeared in your school grounds?

Your chapter starter

Look at the photo on page 118. What do you think happened here ?

Could anyone have stopped it?

How do you think the people felt about it?

Where do you think they've gone ?

Do you think people will ever come back here to live?

A slice through the Earth

You know quite a lot about the outside of the Earth, where you live.
But what's the inside like? You'll find out in this unit.

The Earth's three layers

The Earth is made up of three layers:

1 The crust
This is the layer you live on. It is a thin skin of rock around the Earth, like the skin on an apple (shown here by the thin blue line).

2 The mantle
It forms about half of the Earth. It is made of heavier rock.
The upper mantle is hard. But the rock below it is hot and soft, like soft toffee. It is runny in places.

3 The core
It is a mainly iron, mixed with a little nickel. The **outer core** is liquid. The **inner core** is solid.

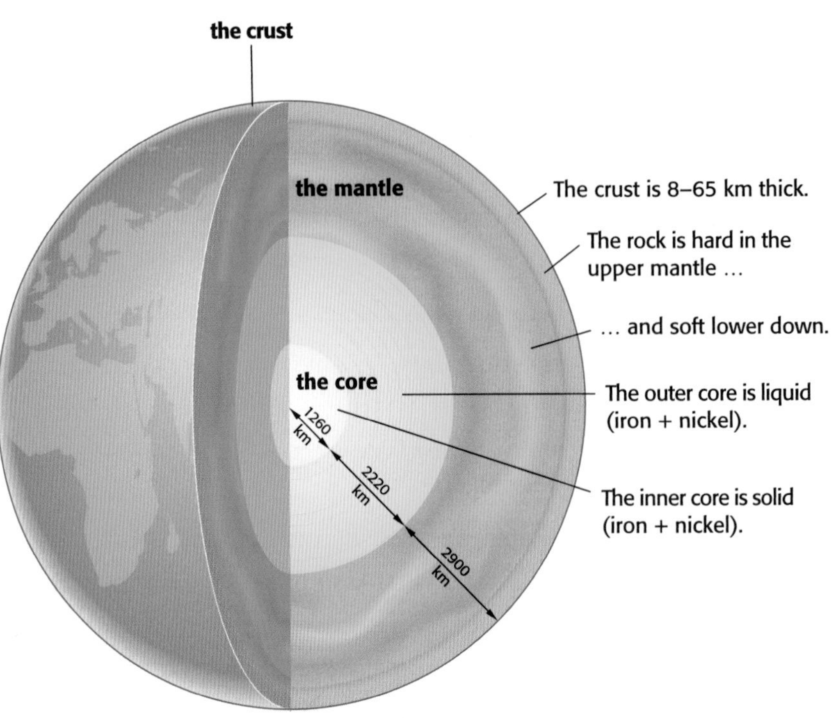

How did the layers form?

Some time after the Earth formed, it got so hot that everything inside it melted. The heavier substances in the liquid sank and the lighter ones rose, making layers. As the Earth cooled, some of the layers hardened.

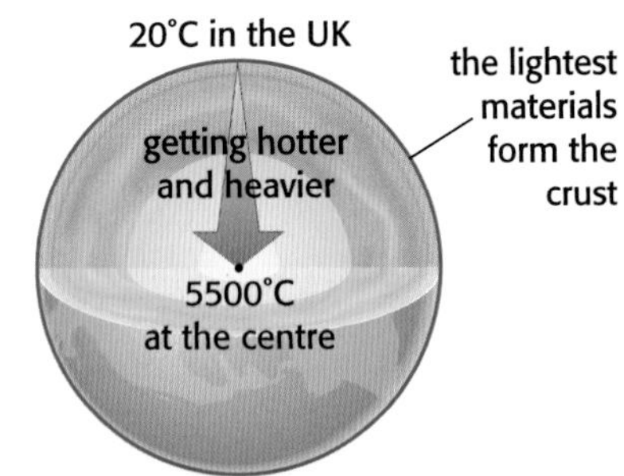

Hot hot hot

It's still very hot inside the Earth. It gets hotter as you go down through it. 200 km down, the rocks are glowing white hot. At the centre of the Earth, the temperature may be around 5500 °C. (We don't know for certain.)

▲ *A bubble of boiling rock reaching the Earth's surface in Hawaii.*

▲ *Several countries have dug holes to find out more about the Earth's crust. The deepest is in Russia – over 12 km !*

More about the crust, and what's below it

This drawing shows part of the crust and mantle.

The crust under the oceans is a thin layer of heavy rock. It is called the **oceanic crust**. The crust that forms the continents is made of lighter rock. We call it the **continental crust**.

Did you know?

- The world's tallest mountain is Mauna Kea in Hawaii.
- It rises 10.23 km from the floor of the Pacific Ocean.

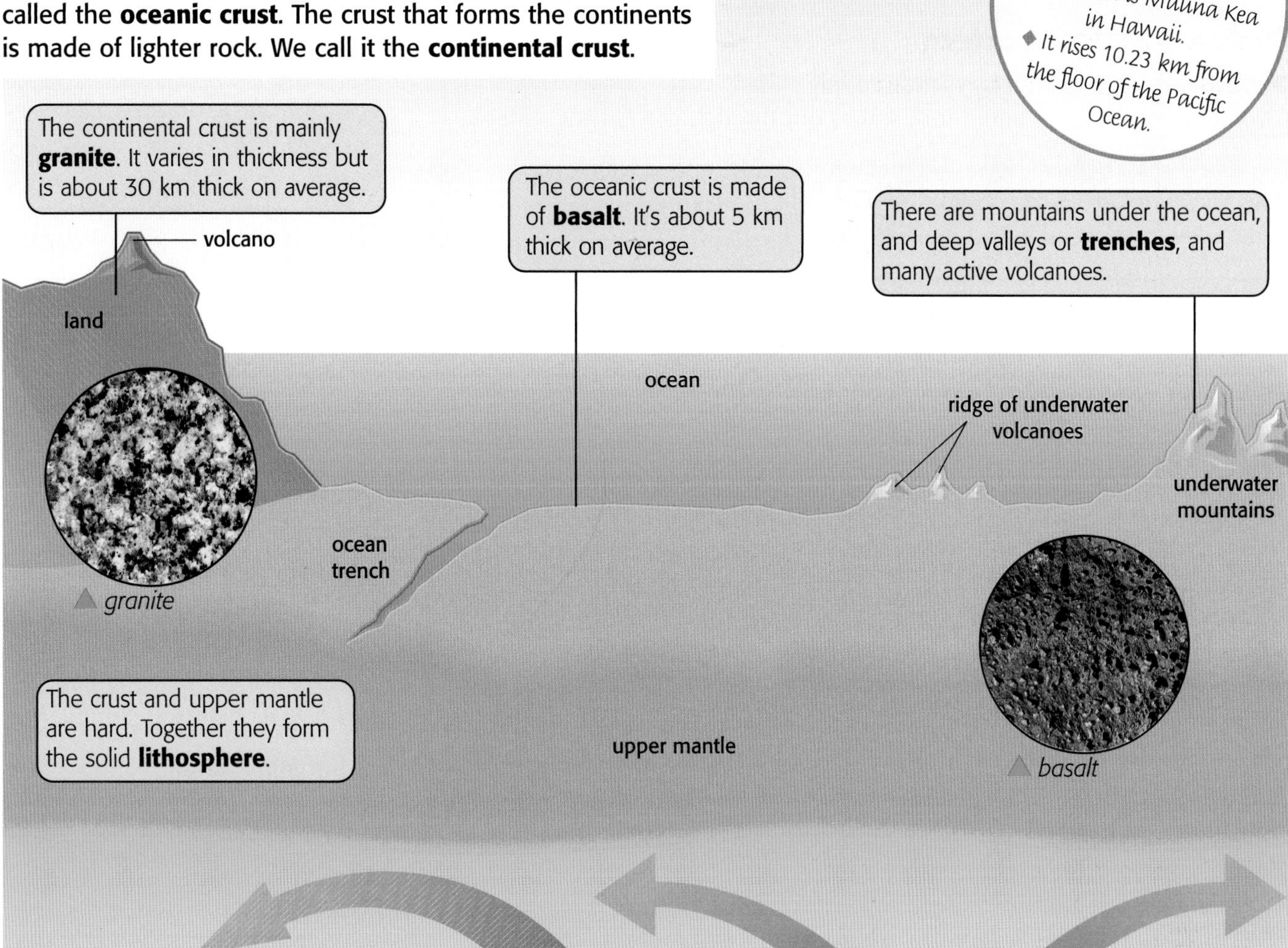

Your turn

1 Make a table like this, and fill it in for the Earth's layers.

Layer	Made of ...	Solid or liquid?	How thick?
crust			
mantle			
core – outer – inner			

2 a What is the Earth's radius, in km, at the thickest part of the crust?

b If you cycle at 20 km an hour, how long will it take you to cycle to the centre of the Earth?

3 Make a larger drawing like this, and complete the labels.

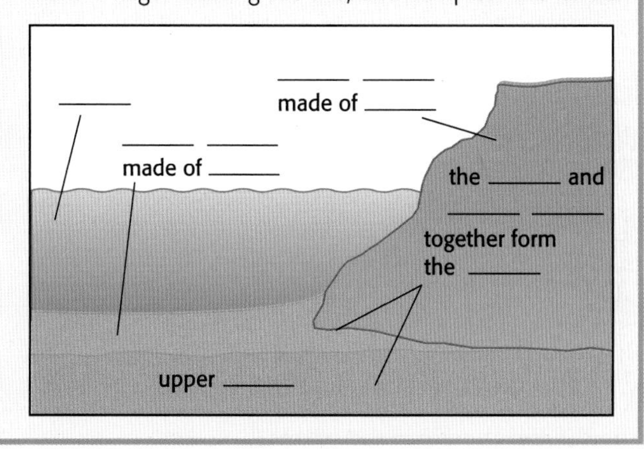

Our cracked Earth

In this unit you'll learn about how the Earth is cracked into huge slabs – and how these are linked with earthquakes and volcanoes.

Did you know?

- *The ring of volcanoes around the Pacific Ocean is known as the Ring of Fire.*

First, a puzzling pattern

An **earthquake** is caused by rock suddenly shifting.
A **volcano** forms when liquid rock spews out through the Earth's surface.

The world map below shows the main earthquake and volcano sites. (This view shows North and South America in the centre.)
Take a good look. Can you see any patterns?

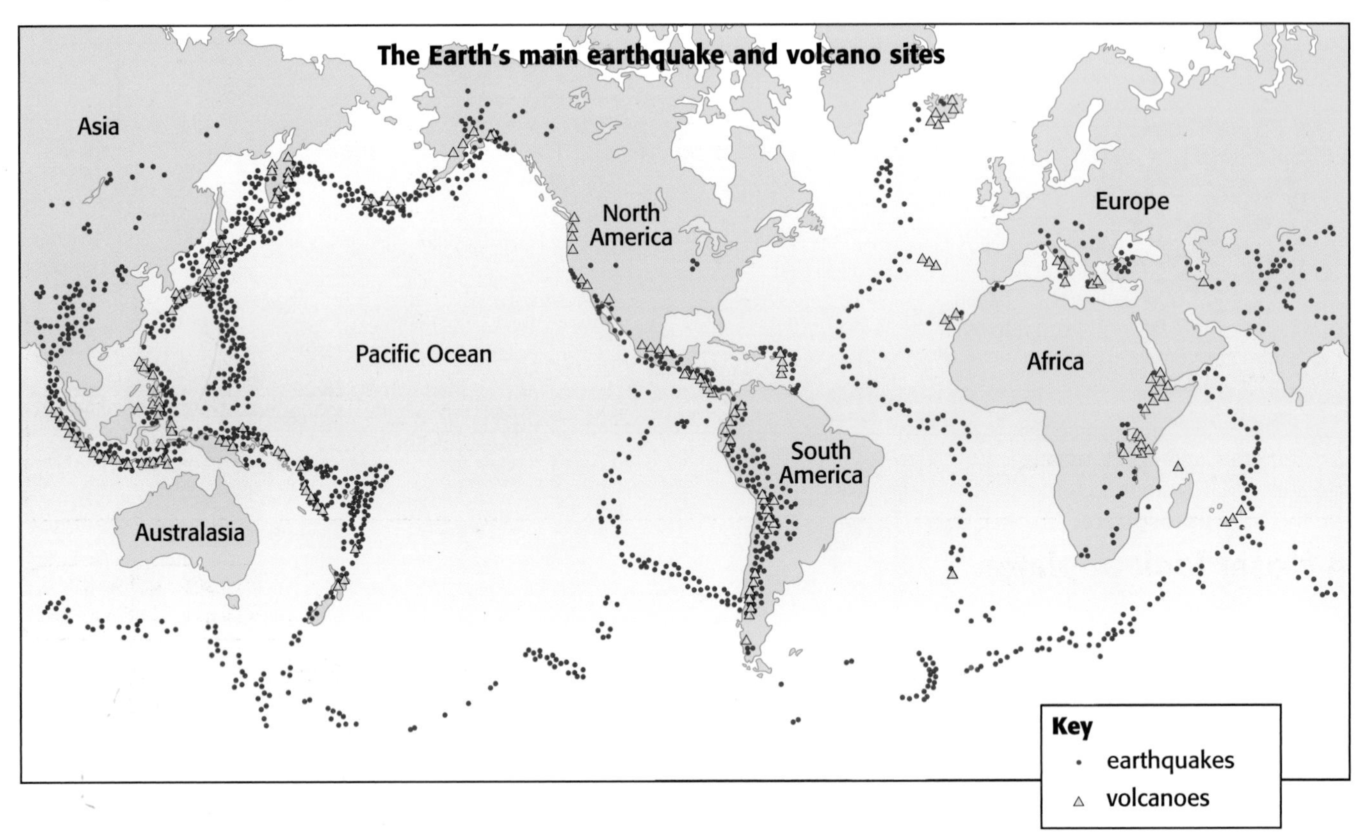

The map shows that:

- Earthquakes and volcanoes don't happen just anywhere. They tend to occur along lines.
- They often occur together.
- They occur in the ocean as well as on land.

Explaining the pattern

The pattern puzzled scientists for years. Then they found the explanation:

- The Earth's surface is cracked into big slabs. (It's like a big cracked eggshell.)
- The slabs are continually moving.
- This movement causes earthquakes, and volcanic eruptions, at the edges of the slabs.

They called the big slabs **plates**.

▲ *Research ships like this one helped scientists solve the plate puzzle. They are used to study the ocean floor.*

The Earth's plates

This map shows the main plates and their names. Some plates carry continents and ocean, others just ocean. They move slowly in different directions.

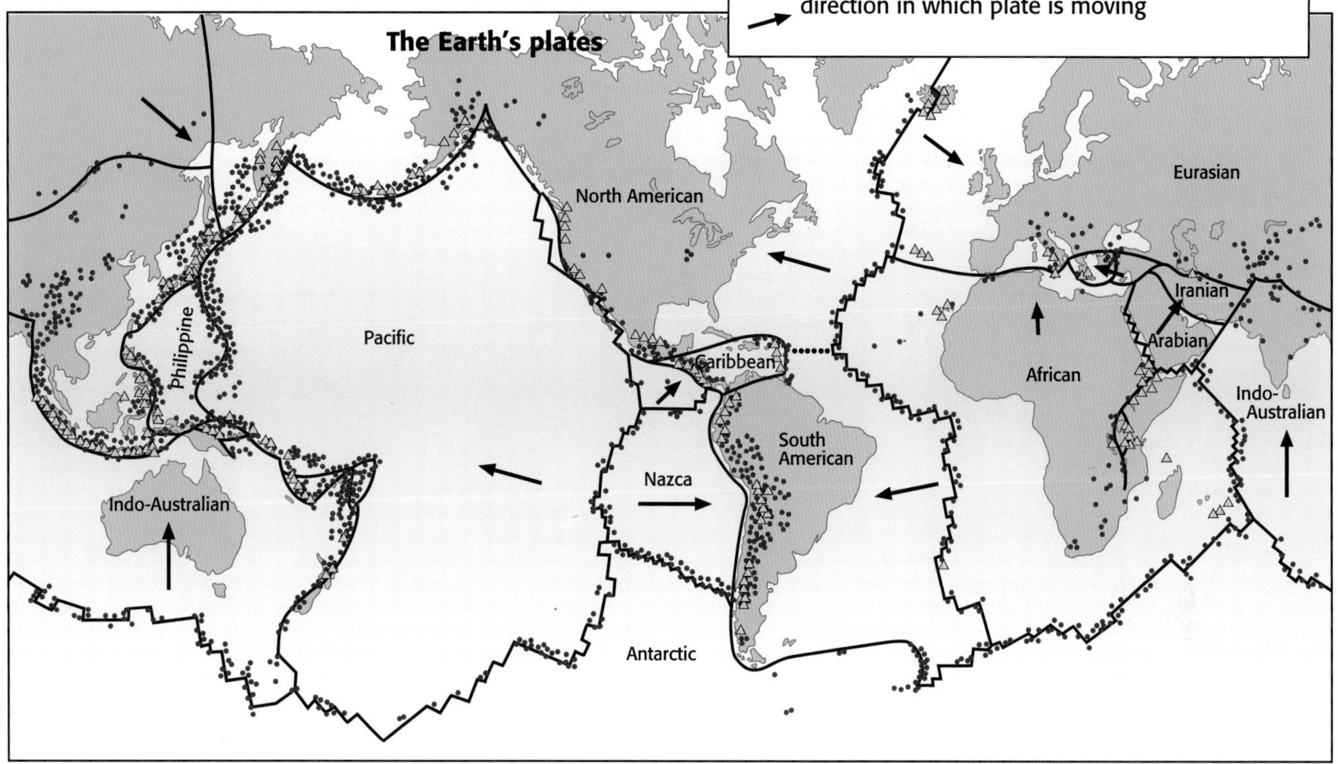

A closer look at plates

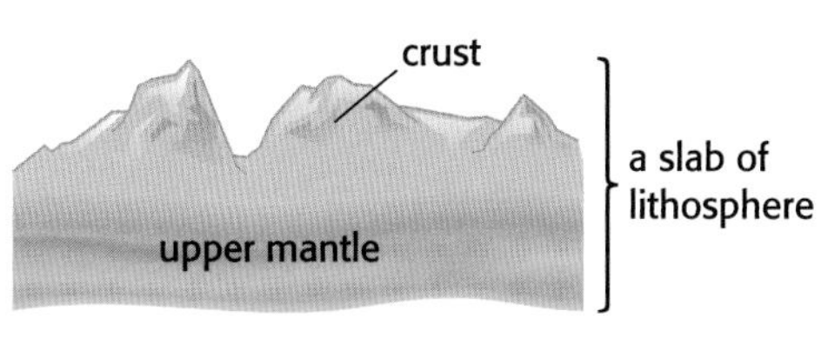

convection current acts like a conveyer belt

convection current

If I just lie still for 10 years I'll move half a metre.

Plates are slabs of the **lithosphere** – Earth's crust and upper mantle. They float on the soft hot rock below.

Plates move because they are dragged along by the powerful hot currents (or **convection currents**) in the soft hot rock.

Plates move just a few cm a year – but it all adds up! For example India has moved 2000 km north in the last 70 million years.

Your turn

1 What is: **a** an earthquake? **b** a volcano? (Check the glossary?)

2 Name:
- **a** the plate you live on
- **b** a plate that is moving away from yours
- **c** a plate that is moving north
- **d** a plate that carries just ocean
- **e** the plate off the west coast of South America
- **f** the plate that's circled by the Ring of Fire.

3 Make a drawing of your own to show what plates are made of, and why they move. Give it a snappy title!

4 Earthquakes and volcanoes form a pattern around the Earth. Using the idea of plates, explain why.

5 The UK has no active volcanoes. Give a reason.

6 A challenge! Suppose our plate starts moving south at 5 cm a year. About how long will it take Newcastle to reach the equator? (Newcastle is about 55° N. A move of 1° south equals 440 km.)

How are the plates moving?

> **Did you know?**
> ◆ London moves 2 cm further from New York every year ...
> ◆ ... because the Atlantic Ocean is getting wider.

In this unit you'll learn how the Earth's plates are moving – and producing earthquakes, volcanoes, and even mountains!

1 Some plates are moving apart

Our plate and the North American plate are moving apart, under the Atlantic Ocean. (Look at the map on page 123.)

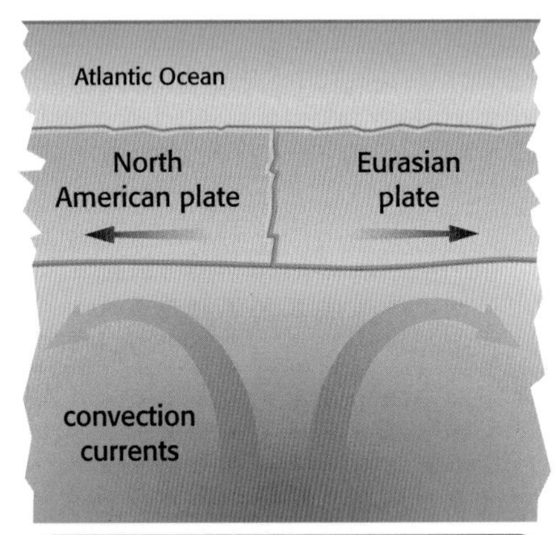

1 The plates are pulled apart by the convection currents in the soft rock below them.

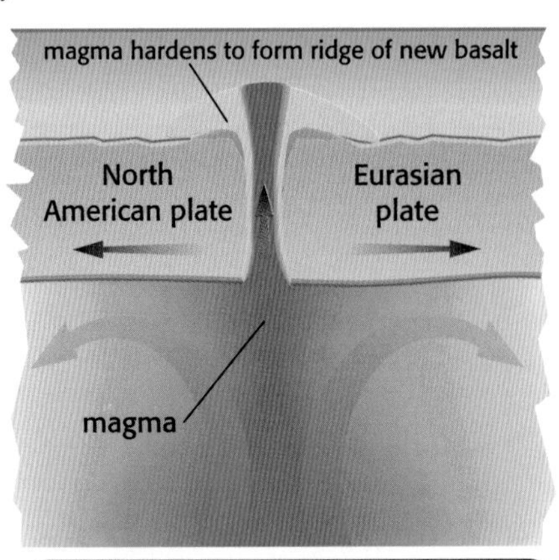

2 Liquid rock or **magma** rises between the plates. It hardens to basalt ...

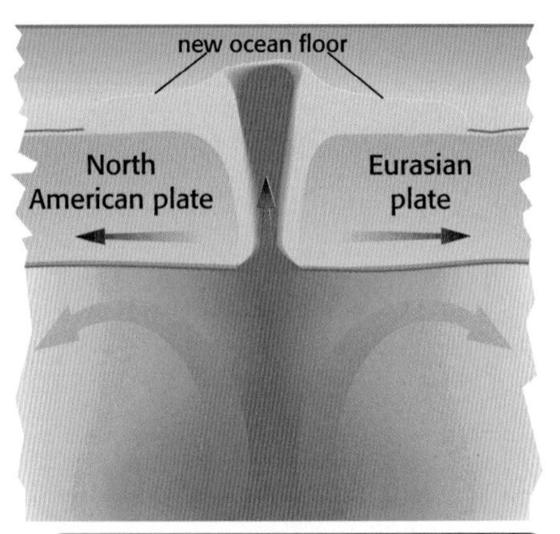

3 ... which forms new ocean floor. So the ocean floor is getting wider – by about 2 cm a year.

The rising magma forms a line of volcanoes under the ocean. The heavy moving plates cause earthquakes too. So, where plates are moving apart, you get earthquakes, and volcanoes, and new ocean floor being formed.

2 Some plates are pushing into each other

The Nazca plate and the South American plate are pushing into each other, just off the west coast of South America. (Look at the map on page 123.)

The result is earthquakes and volcanoes.

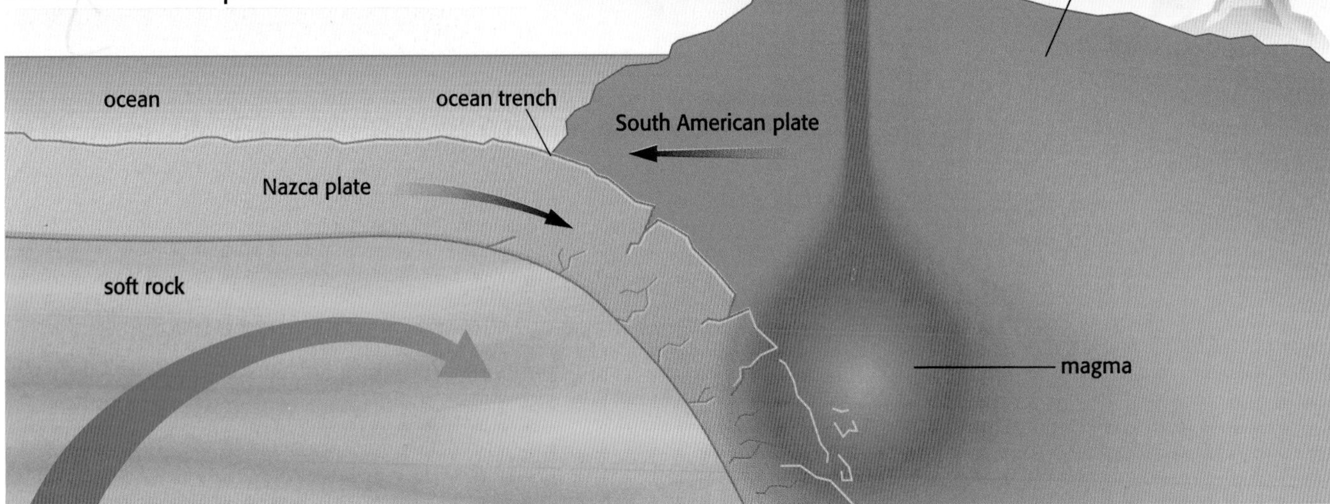

1 The Nazca plate is heavier. (Oceanic crust is heavier.) So it gets pushed under at an ocean trench.

2 The rock jolts and grinds its way down, causing earthquakes. At the same time ...

3 ... it heats up. Some rock melts, and forces its way up through the Andes to form a volcano.

When pushing makes mountains
This simplified drawing shows how the Indo-Australian and Eurasian plates are pushing into each other.

As a result of the pushing, rock has got squashed up to form mountains: the Himalayas.

The plates are still pushing. So the Himalayas are still growing – and China (on the Eurasian plate) gets lots of earthquakes.

The Himalayas are called **fold mountains**. Can you see why?

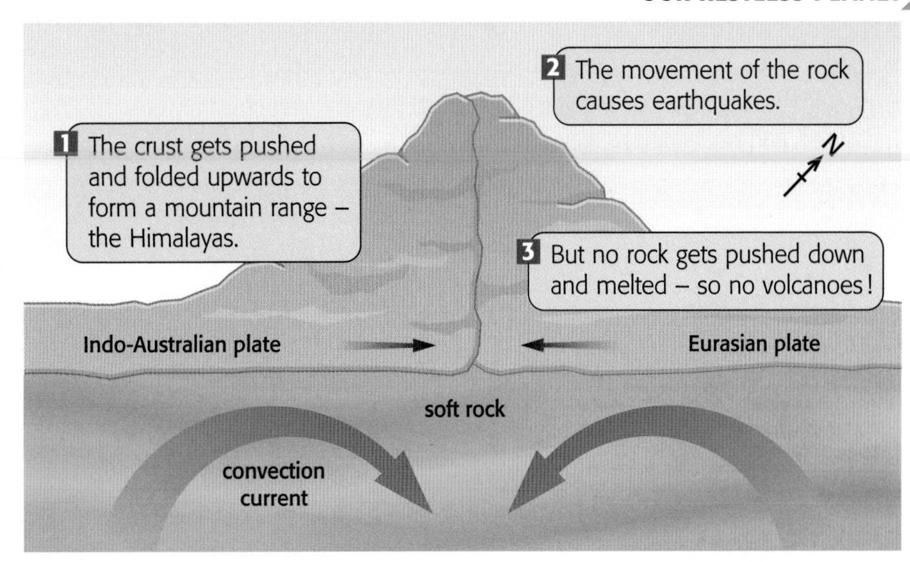

3 Some plates are sliding past each other

The Pacific plate is sliding past the North American plate. (Look at the map on page 123.)

Both move in the same direction, but the Pacific plate is moving faster.

The result is earthquakes now and then – but no volcanoes!

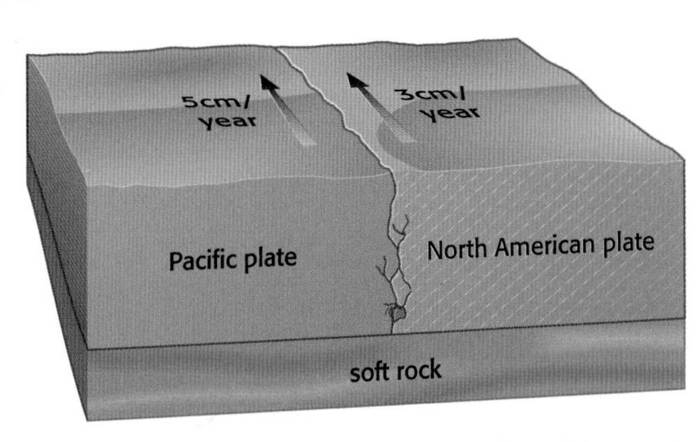

1 Parts of the plates get stuck, then lurch free. This causes earthquakes.

2 But no rock gets pushed down and melted – so no volcanoes.

Your turn

1 The photo on the right shows the floor of the Atlantic Ocean. The grey ridge lies along plate edges.
 a Name the plates that lie on each side of the ridge.
 b What is the ridge made of?
 c Explain what is happening along the ridge.
 d Do you think earthquakes occur there? Explain.
 e Where else might you find a ridge like this?

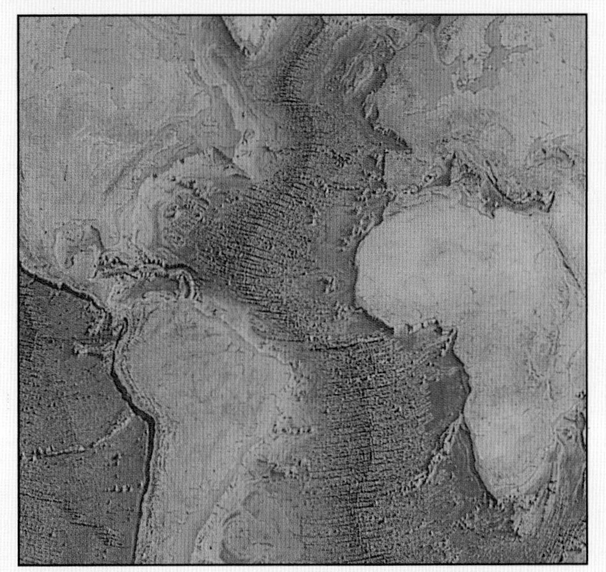

2

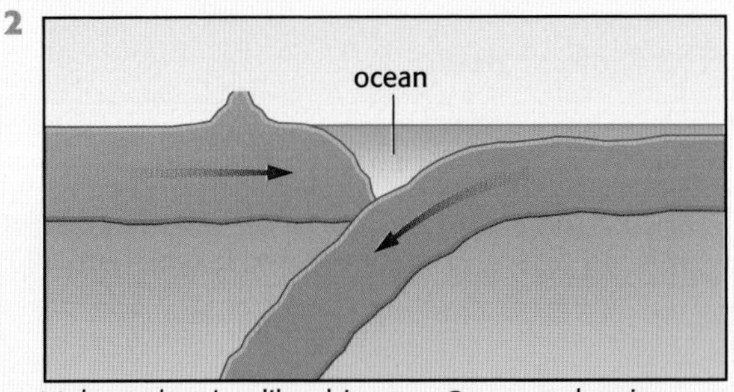

Make a drawing like this one. On your drawing:
 a label the ocean plate, the continental plate and a volcano.
 b mark in melted rock that feeds the volcano.
 c mark in and label an earthquake site.

3 Now, using the maps on pages 123 and 140–141, explain why:
 a Peru has earthquakes and volcanoes
 b Iran has fold mountains
 c Italy has earthquakes and volcanoes
 d Japan has earthquakes and volcanoes.

Earthquakes

Earthquakes can kill. In this unit you'll learn what they are, and how they are measured, and what damage they do.

What is an earthquake?

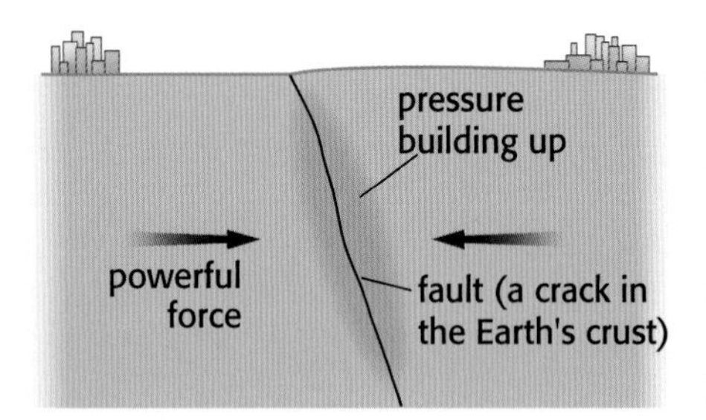

Imagine powerful forces pushing two huge masses of rock into each other. The rock stores up the pressure as **strain energy**.

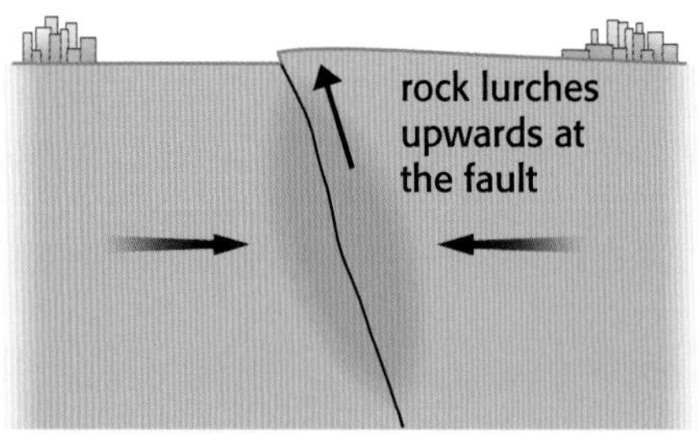

But suddenly, the pressure gets too much. One mass of rock gives way, slipping upwards. The stored energy is released in waves …

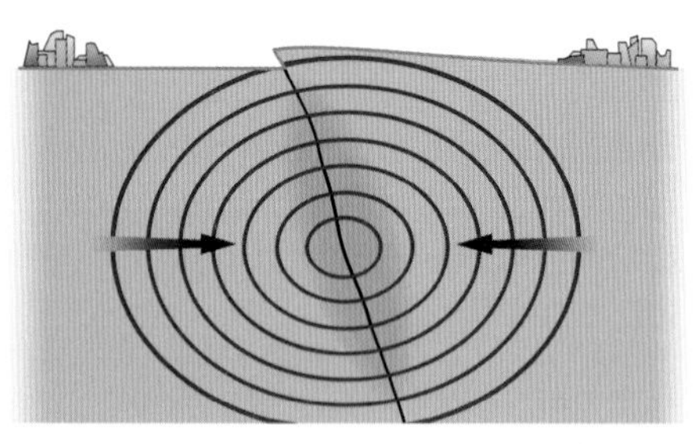

… called **seismic waves**. These travel through the Earth in all directions, shaking everything. The shaking is called an **earthquake**.

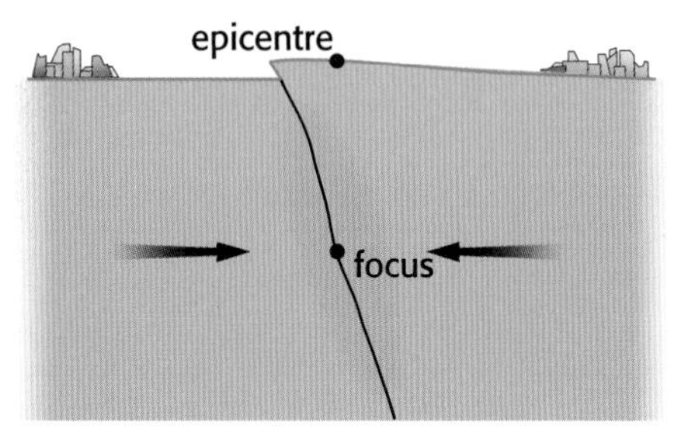

The **focus** of the earthquake is the point where the waves started. The **epicentre** is the point directly above it on the Earth's surface.

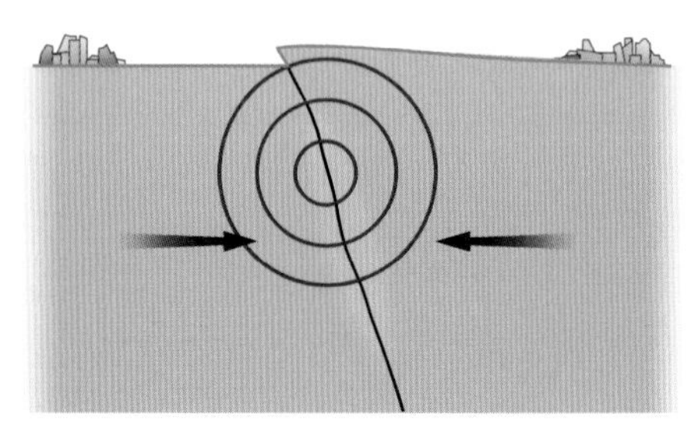

As the rock settles into its new position, there will be lots of smaller earthquakes called **aftershocks**.

Seismic waves get weaker as they travel. Even so, a large earthquake can still be detected thousands of kilometres away!

Any sudden large rock movement will cause an earthquake. That's why there are so many earthquakes along plate edges. But even the collapse of an old mine shaft, or an underground explosion, can cause a small earthquake.

How big?

- Earthquakes are measured using machines called **seismometers**. These record the shaking as waves on a graph.
- From the graph, scientists can tell how much energy the earthquake gave out.
- The amount of energy an earthquake gives out is called its **magnitude**.
- We show it on the **Richter scale**. (On the right.)
- An increase of 1 on this scale means the shaking is 10 times greater, and about 30 times more **energy** is given out. (And that means a lot more damage.)

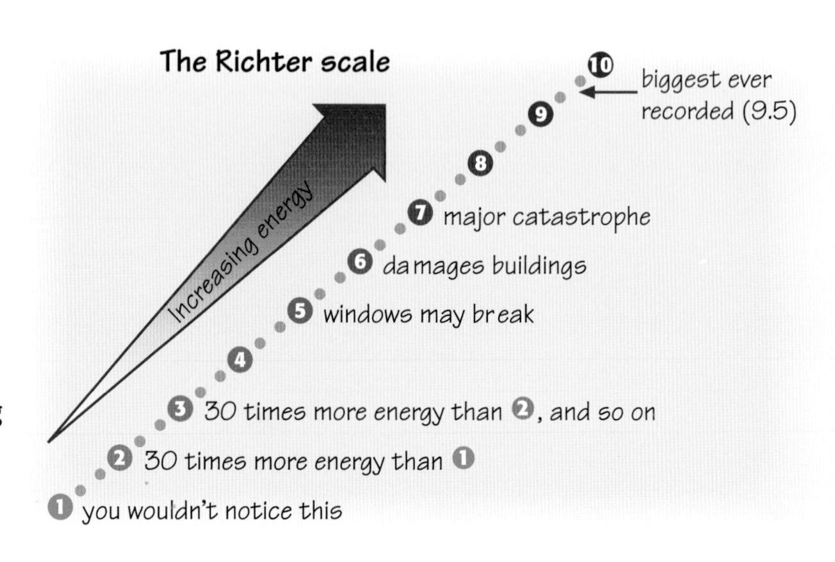

The damage it can do

An earthquake shakes the ground, which then shakes everything on it. So …

So earthquakes can destroy villages, towns and cities. You will see an example in the next unit.

Your turn

1 Why do earthquakes happen so suddenly?

2 Explain in your own words what these earthquake terms mean. Use complete sentences.
  a seismic waves b focus c epicentre
  d magnitude e seismometer

3 You are one of the people in the photo above. (Look carefully!) Describe what you see around you, in about 100 words.

4 Look at the earthquake diagram on the right.
  a Will the shaking be stronger at A, or at B? Explain.
  b Will the damage be greater at A, or at B? Why?
  c Will an earthquake of magnitude 7 do more damage than this one, or less? Why?
  d About how many times more energy will an earthquake of magnitude 7 give out, than this one?
  e An earthquake can occur at any time of day. When might an earthquake do more harm at B?
    i at 5 am ii at 10.30 am
    Explain your answer.

5 The largest earthquake ever recorded was in 1960, *in the ocean off the coast of Chile*. It measured 9.5 on the Richter scale. Use the maps on pages 123 and 140–141 to help you answer these questions.
  a What do you think caused the earthquake?
  b It left 2 million people homeless in Chile. See if you can explain how it managed to do that.
  c 22 hours later, it caused 200 people to drown on the east coast of Japan. How did it do that?

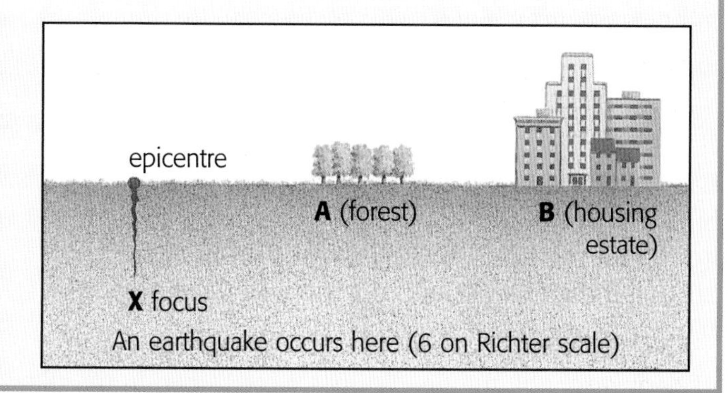

9.5 Earthquake in Pakistan

In this unit you'll learn about a big earthquake in Pakistan, and what caused it, and why it did so much damage.

Lucky Yasin

Yasin is a lucky boy. He is alive. His friends are dead.

Yesterday was Saturday, 8 October. He should have been at school, as usual. But he had a cold, so his mum kept him home. And that saved him.

At 8.50 in the morning, a massive earthquake struck Balakot. Yasin was in the main room, reading. His mum was in the courtyard with his little sister, chatting to a neighbour. Suddenly, there was a roar like thunder. The ground shook. He dashed out the door. Just in time. With a sickening crack, the house collapsed behind him.

It was terrible. First they ran hither and thither, in shock and panic. They met people crying, and carrying wounded children, and begging for help. They ran to his uncle Saleem's house. It was nothing but rubble, and silence.

Next they sat on the patch of open ground at the corner, cold and fearful. Everywhere they looked, homes lay in ruins. Every so often the ground shook again, and they clutched each other.

Then Yasin thought of school. He ran all the way, weaving through the debris. When he got there, he could hardly take it in. No school. The building had been flattened. Around the edges were frantic parents, tugging at slabs of concrete. He went to the corner where his classroom had been. There was no sound from the rubble. He too began to search.

And then he saw the red tin box that belonged to Latif, his best friend. It was crushed flat. He picked it up, and bowed his head, and wept.

Based on news reports, 9 and 10 October 2005

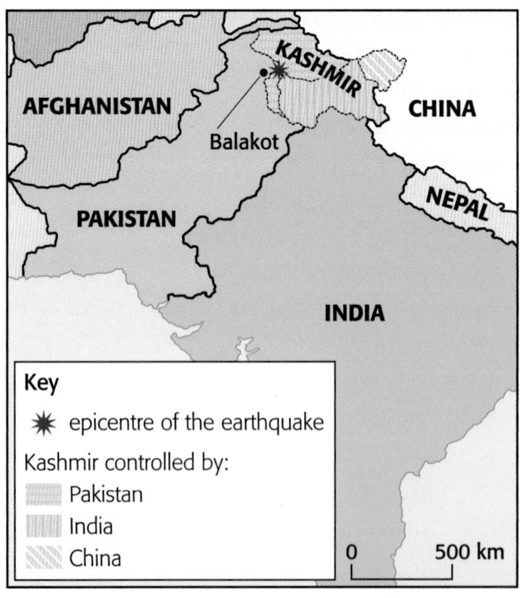

▲ *All that's left of her home in Balakot.*

Widespread damage

The map above shows where the earthquake struck.

It affected a large area. It even caused deaths in Afghanistan. But the most damage was in northern Pakistan and Kashmir.

In Kashmir, dozens of villages high up in the mountains were destroyed. Poor roads, and landslides, made it hard to reach them. In some villages, survivors waited for help for many weeks, with little food or shelter, in bitter cold. Many died while waiting.

Balakot, Yasin's town, had been a busy resort for tourists and hikers. 35 000 people had lived there. But it was only 20 km from the epicentre. After the earthquake it lay in ruins. 80% of the buildings had collapsed – including several schools. Most of the students were killed.

At Yasin's school, the dead pupils were buried in the playground.

The earthquake factfile

date	Saturday 8 October 2005
time	8.50 am
magnitude	7.6 on the Richter scale
epicentre	inside Pakistan-controlled Kashmir, about 20 km from Balakot
damage	over 74 000 people killed, over 106 000 injured, and over 3.3 million people left homeless, in Kashmir and northern Pakistan
financial cost	nearly £3 billion

What caused the earthquake?

Like most big earthquakes, this one was caused by plate movements.

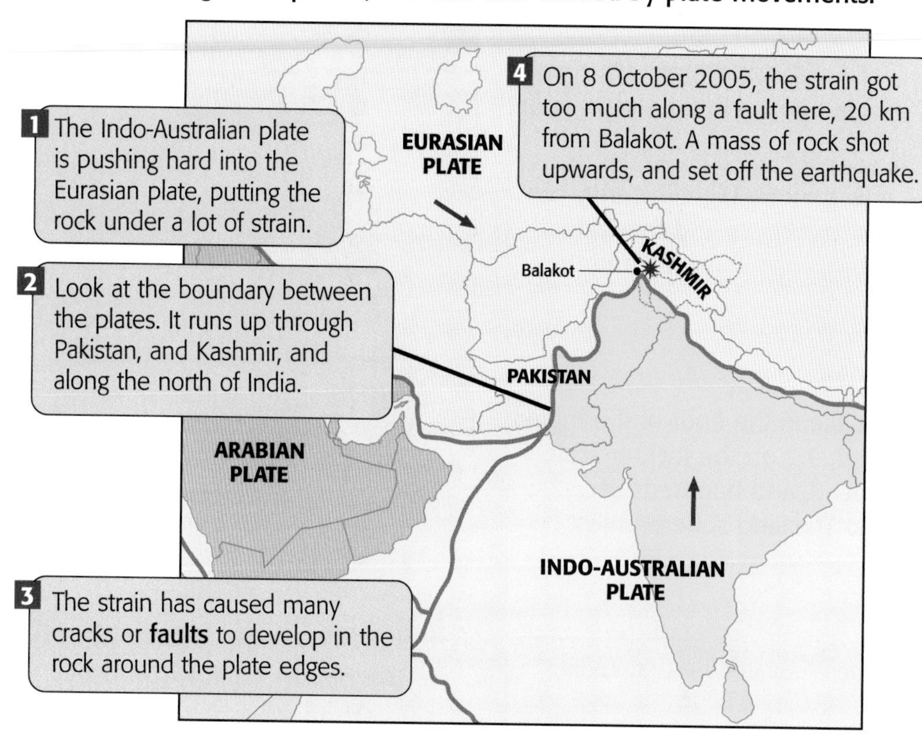

▲ *Balakot from the air, the day after the earthquake.*

▲ *Rescue workers search in the rubble of a school in Balakot.*

Why was there so much damage?

Earthquakes don't kill. Collapsing buildings do. And that was the problem.

The area has a very high risk of earthquakes. But most buildings, even the hospitals and schools, were not built to cope with them. Many people in the villages had built their own houses.

What next for Balakot?

Balakot is being rebuilt – but in a safer spot, 20 km south of the old town. It will be a modern town, with quake-proof buildings. There will be space for it to grow, and in time it could hold up to 2.5 million people.

The UK, France, and Libya are all helping to pay for the new town. Work is due to finish in 2010.

Your turn

1 Where is Pakistan? Say what continent it is on, and which countries border it. (Page 141 may help.)

2 The epicentre of the earthquake was in Kashmir. But people called it **the Pakistan earthquake**. Explain this.

3 What caused the earthquake? Explain in simple words, as if to a 7-year-old.

4 Look at the photo on page 128. Imagine you are that woman. Write a diary entry describing what happened to your home and family on 8 October 2005, and how you feel about it.

5 Look at this person's message. Do you agree with it? Give your reasons.

6 Experts say there will be more big earthquakes in northern Pakistan and India. Why do they think so?

7 You are the minister in charge of town planning and building, for Pakistan. You worry about earthquakes. What will you do to protect people? Write your plans as bullet points, in order of importance.

Tsunami!

In this unit you'll learn about tsunami, and the damage they do.

What is a tsunami?

A tsunami is a series of waves, set off by an earthquake in the ocean floor. The waves spread in all directions.

Out in the deep ocean the waves may be only a metre high. But they can travel at over 700 km an hour. As they approach a coast, and the water gets shallower, they get slower, and taller. When they finally hit land they can be up to 30m high. And they are deadly.

▲ *People flee as the tsunami slams onto Raya island in Thailand.*

Tsunami in the Indian Ocean, 2004

On 26 December 2004, there was an earthquake in the floor of the Indian Ocean. It was the second largest ever recorded: 9.2 on the Richter scale. It set off a tsunami that left 230 000 people dead, and hundreds of thousands homeless. Indonesia, Sri Lanka and Thailand suffered most. Follow the numbers below for more.

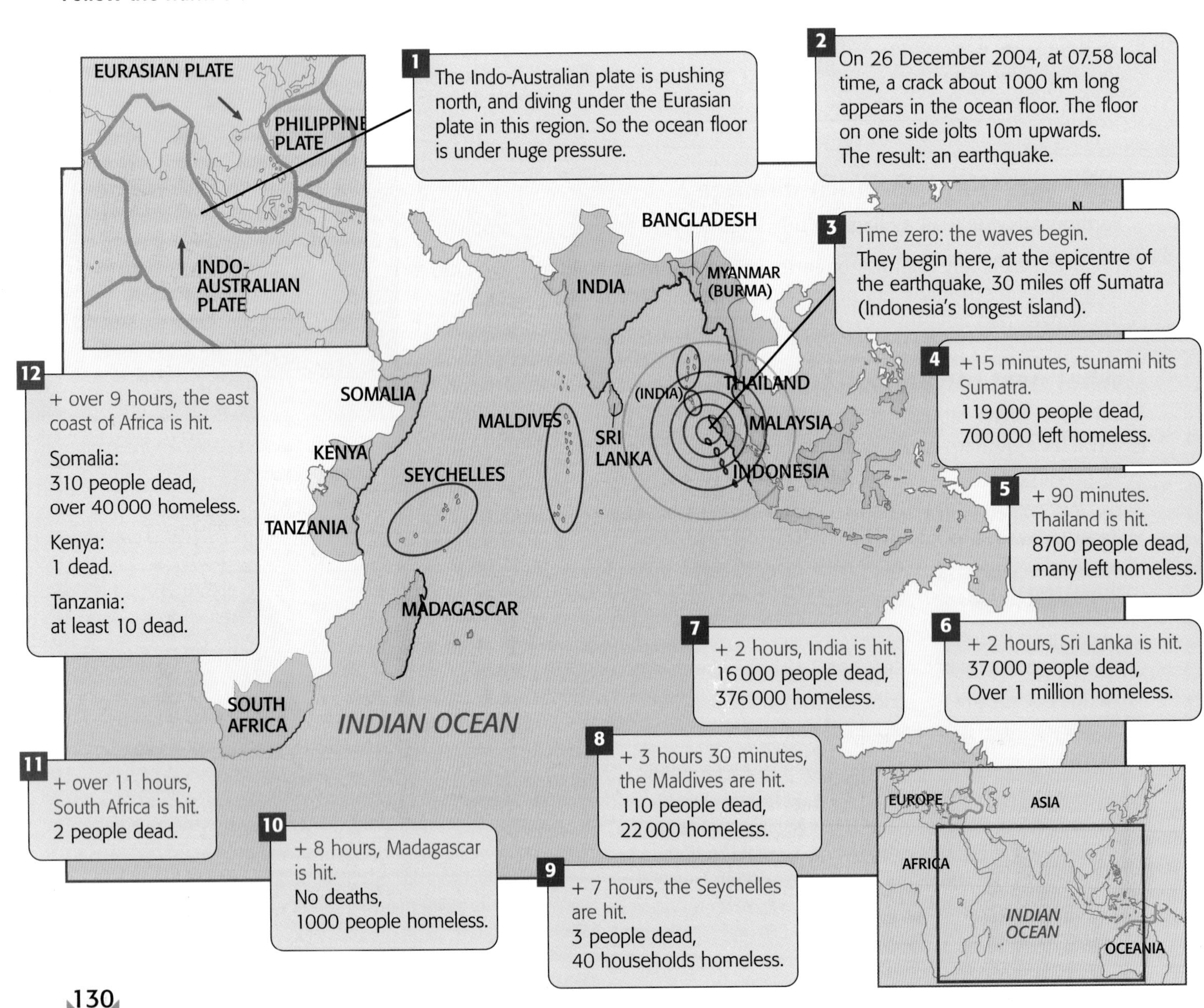

▲ *As a tsunami nears the coast, it sucks up water, exposing the ocean floor. This satellite image shows the water being dragged away, at the resort of Kalutara in Sri Lanka.*

▲ *The same resort, after the tsunami has struck. The water churns and recedes, leaving destruction behind it. Both images were taken on 26 December 2004.*

The day they will never forget

Banda Aceh, Indonesia: I took an early ferry. I thought it was bouncing a bit, but that did not worry me. After an hour we got to Banda Aceh. I could not believe my eyes. The fishermen's homes along the water had gone. In the town, there were fishing boats on roof tops, and taxis stuck in trees. There were people sobbing, and corpses lying everywhere.

Telwatta, Sri Lanka: I was on the coast train, going see my family. Suddenly the train stopped. The sea started to pour in, very fast. The train turned over and over. I was trapped in there for nearly an hour, half drowned. But I'm lucky. They say there were 1500 passengers, and 800 of them died.

Khao Lak, Thailand: There was a hissing noise, and all the water along the beach got sucked out to sea. There were lots of fish left flapping on the ground. Children ran to look at them. Then there was a noise like thunder, and we saw a giant wave coming. The children had no chance.

Cuddalore, India: My two sons were playing cricket on the beach, with about 40 other children. I could hear them shouting and cheering. Then I looked down from the window and saw huge waves coming in, about 10 or 12 metres high. I froze. The water churned round and round. And then it sped out to sea, dragging them with it.

(Adapted from news reports, end December, 2004)

▲ *Three days after the tsunami, on the island of Phi Phi in Thailand.*

Did you know?

◆ *In the Boxing Day tsunami, waves travelled nearly 5000 km from the epicentre to Africa.*

Your turn

1 What causes a tsunami?

2 Try to explain these facts about the Boxing Day tsunami.
  a The tsunami reached more than a dozen countries.
  b The tsunami arrived at each at a different time.
  c Indonesia suffered much greater loss than Somalia.
  d People out at sea were not aware of the tsunami.
  e The earthquake was detected in the Philippines, but no tsunami reached there. (Page 141?)

3 People all over the world gave money to help the devastated places. Why do people do this?

4 Before the Boxing Day tsunami, there was a tsunami warning system in the Pacific Ocean (see pages 140–141) but none in the Indian Ocean. Suggest a reason.

5 It's your job to invent a tsunami warning system for the Indian Ocean. It must warn countries in good time when a tsunami is on the way. Tell us your ideas!

9.7 Volcanoes

In this unit you'll learn what volcanoes are, and what damage an eruption can do.

What's a volcano?

A **volcano** is where liquid rock or **magma** shoots out or **erupts** through the ground. Above ground, the liquid rock is called **lava**.

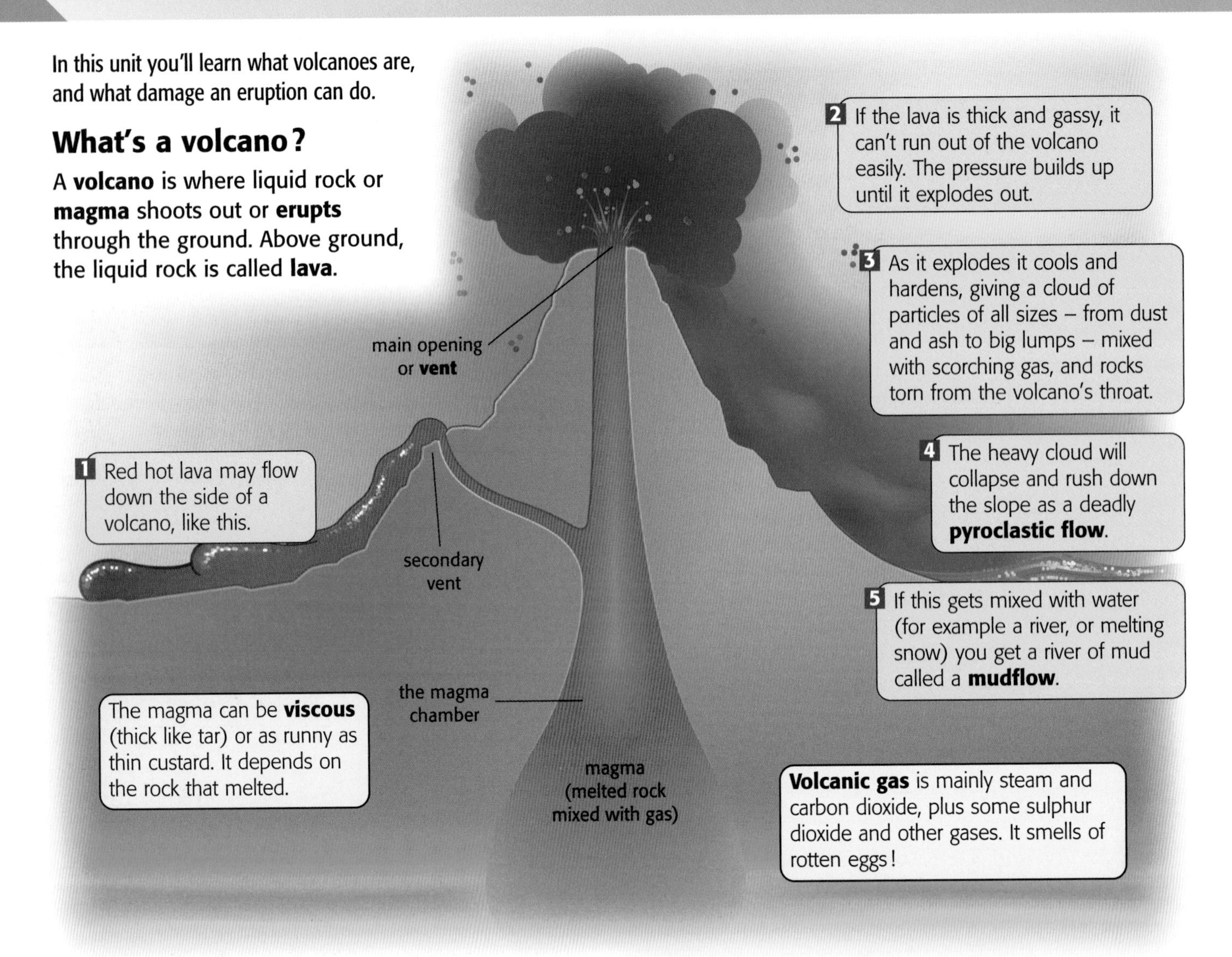

Viscous gassy magma is the most dangerous kind. It builds up inside the volcano. Then the gas propels it out in an explosion.

▲ *An eruption of runny lava in Hawaii.*

▲ *A small explosion of steam, gas and ash from Mt St Helens (USA). The hollow around the vent is called a* ***crater****.*

What damage can eruptions do?

This photo was taken on the island of Montserrat. See next page for more.

Your turn

1 What is: **a** magma? **b** lava?

2 Make a larger copy of this drawing. Then colour it in and add the missing labels.

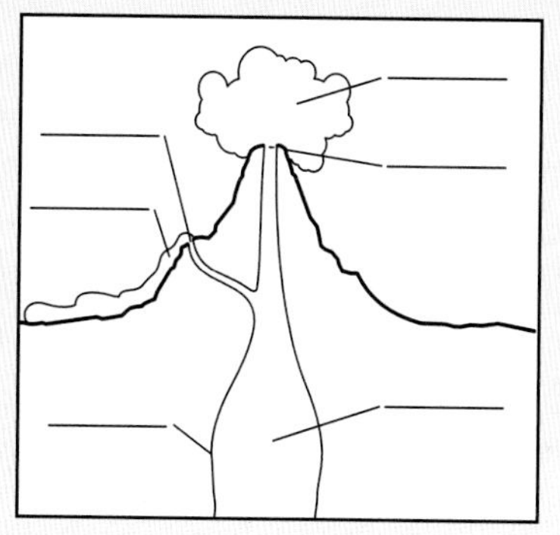

3 Look at the photo above. What do you think happened to:
a the roof of the church? **b** the trees near the church?

4 An active volcano can produce:

showers of ash a pyroclastic flow a lava flow
plumes of dust volcanic gases

a List them in order of danger, starting with what you think is the most dangerous one.
b Beside each item in your list, say what harm it does.

5 You were there when Mount Pinatubo in the Philippines erupted, in 1991. You took the photo below. E-mail your friend in New York telling him what you saw before you took it – and what happened next.

Montserrat: living with an active volcano

In this unit you will learn how an erupting volcano has changed a Caribbean island forever.

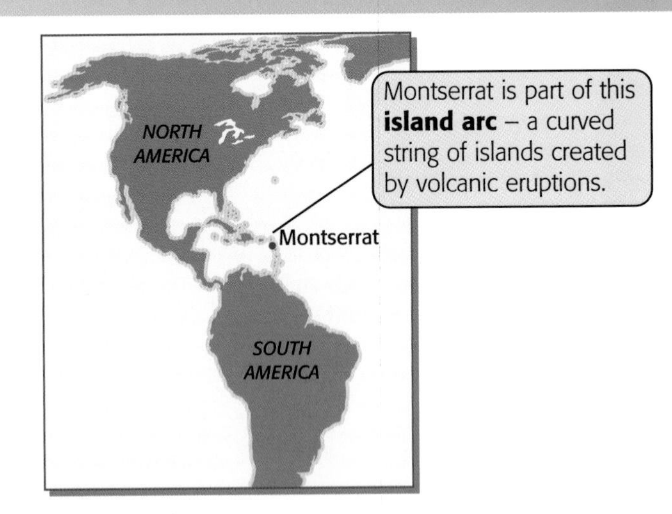

An island paradise

At the start of 1995, 11 000 people lived on the island of Montserrat in the Caribbean. Some were farmers. Some worked in a few small factories. But most depended on the tourists who came to enjoy this island paradise.

Then, on 18 July 1995, life began to change, forever. The volcano, asleep for nearly 400 years, began to waken.

The volcano awakens

The first signs were rumbling noises, and showers of ash, and a strong smell of sulphur. The government acted quickly. It called in **vulcanologists** (volcano scientists) to check or **monitor** the volcano, and made plans to move people to safety.

That was in 1995. Twelve years later, in 2007, the volcano was still busy!

- Over the years, it has blasted out clouds of dust and ash that turned the sky black.
- It has grown domes full of lava that glowed at night before exploding.
- Many pyroclastic flows have raced down the slopes. Some have turned rivers into mudflows.

The vulcanologists watch it night and day. But they can't predict when it will go back to sleep again.

▼ *Montserrat from the air. Spot the volcano!*

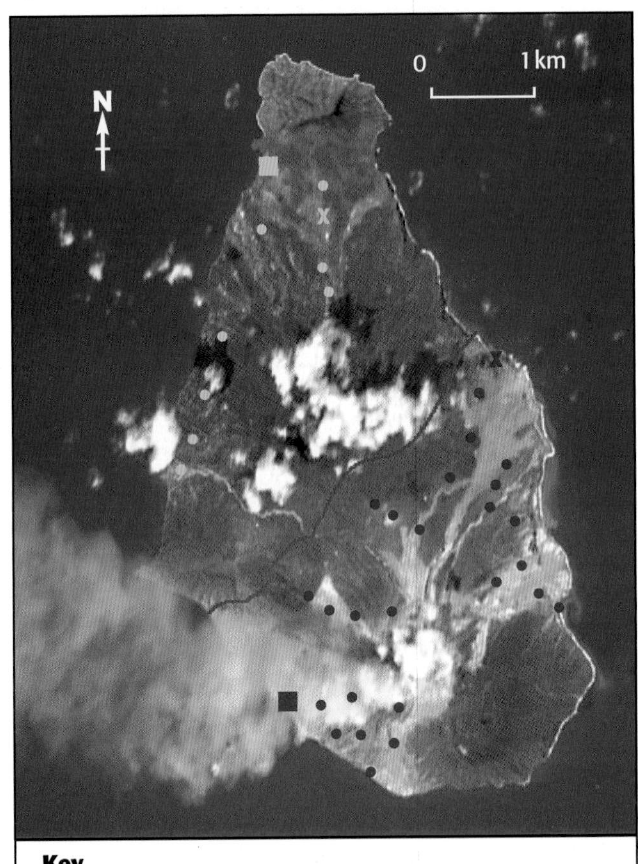

◀ *Just another pyroclastic flow on Montserrat.*

What's causing the eruptions?

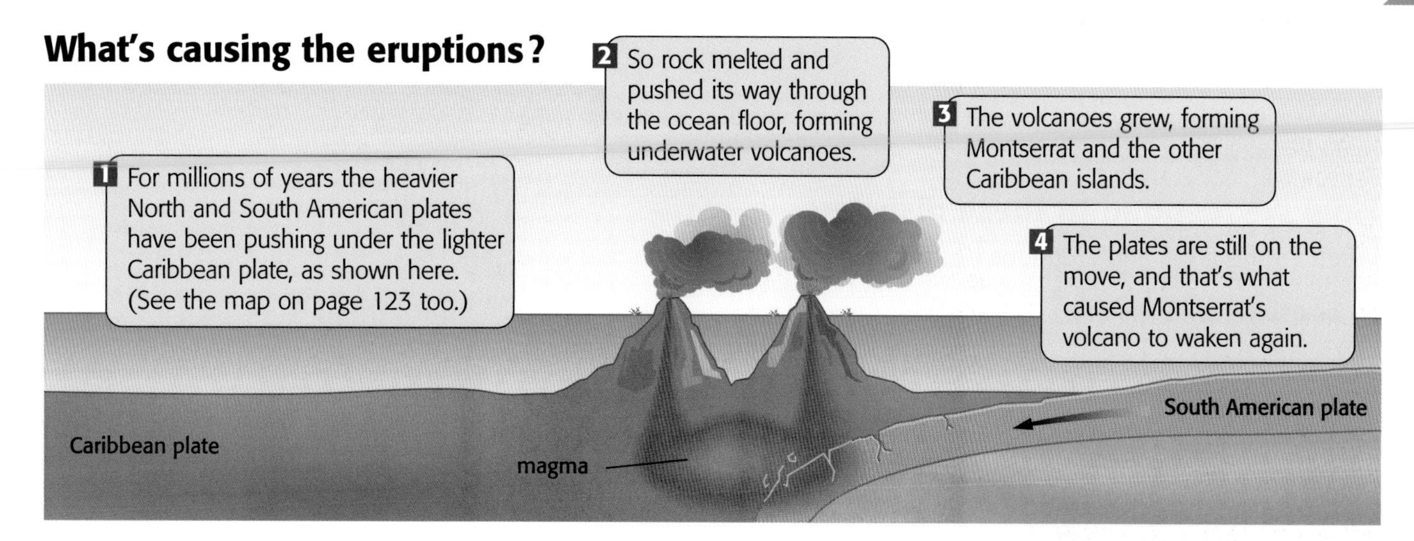

People on the move

As the volcano grew restless, people were moved from the south of the island. Some went to the 'safe' area in the north, to stay with friends or in shelters. Some went to other Caribbean islands, or to relatives abroad. By April 1996 the south of the island was empty.

But some could not keep away. On 25 June 1997, pyroclastic flows killed 19 people who had crept back to work on their farms.

Life goes on

Today, only 4500 people are left on Montserrat, in the north of the island. The south is still out of bounds. (If you are caught there you will be fined and may even be sent to prison.)

There is not much farming now, since so much of the land is ruined. People depend on grants from the UK and the European Union. But they are trying to attract tourists again. A new airport was opened in 2005. A new capital is being built around the small port of Little Bay.

But people can't forget about the volcano. Every so often a dark plume in the sky, or a shower of ash, or a rattle of pebbles, reminds them.

▲ *Out of action… forever?*

Your turn

1 Explain in your own words why the volcano on Montserrat is erupting.

2 Look at the photo on page 118. It shows Plymouth, the capital of Montserrat, destroyed by the volcano. You used to live there. Write a letter to your cousin in Burnley describing what Plymouth looks like now.

3 How will the eruptions on Montserrat have affected:
a farmers? b hotel owners? c taxi drivers?

4 Montserrat hopes to attact tourists again – as a volcano island! You are in charge of tourism.
a Draw a sketch map of the island, showing the volcano, the new airport, and the safe zone.
b Mark in where you would put a new tourist hotel.
c What activities will you lay on for tourists?
d How will you make sure the tourists are safe?
e What kind of souvenirs will you sell them?
f Make up a slogan to attract tourists to the island.

5 Montserrat has received over £200 million in aid, since the volcano awoke. Much of this was from the UK. (Montserrat is a British Overseas Territory.) Some people think the island should just be closed down.
a Give some arguments in favour of this.
b Give some arguments against it.
c If you had to make the final decision, what other information would you want?

Coping with earthquakes and eruptions

In this unit you'll learn how we cope with earthquakes and eruptions – and why some countries find it harder than others.

How we respond to these disasters

When earthquakes and eruptions destroy places, we respond in two ways.

1 Short-term response

First, we quickly try to help the survivors.

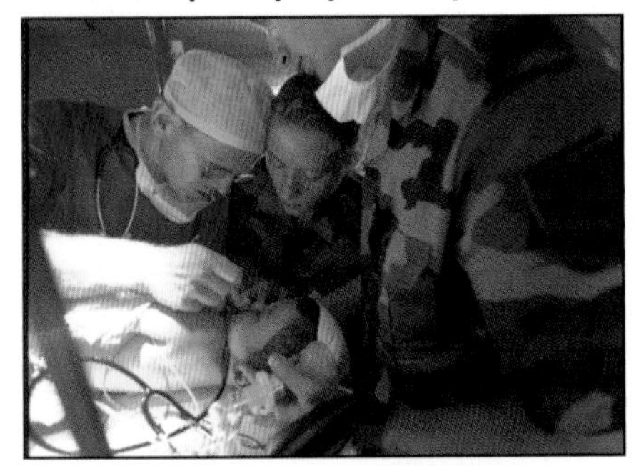

Doctors, nurses, firemen and the army rush there. Medical tents are set up. Aid agencies like Oxfam and the Red Cross arrive.

Tents, food, water and clothing are given to the people who have lost their homes. (These may be gifts from other countries.)

Ordinary people like you and me give money, to help the survivors get food and shelter, and rebuild their lives and homes.

2 Long-term response

Then we try to prevent disasters like this happening in the future.

We can't stop earthquakes and eruptions. But geologists can at least identify the areas at risk.

Then we can make plans to protect the people living in those areas. Some of the plans …

… can be put into action right away. For example build quake-proof homes like these.

Other plans will be put into action the next time there's an emergency.

Meanwhile, scientists can monitor the areas at risk, and try to predict when earthquakes or eruptions …

… will happen, so that they can warn people in time to move to a safe place.

Your turn

1 **A–G** below match words or terms used on page 136. For each, you have to pick out the matching word or term. Set out your answers like this:
B for the months and years ahead = long–t…

A for the next days and weeks
B for the months and years ahead
C a sudden dangerous situation
D to observe and check and take measurements
E to tell in advance when something will happen
F they study rocks (and earthquakes)
G they collect money to help people in disaster areas, and supply them with things like clean water

The glossary might help!

2 There is a big **earthquake** in a city in South America. 80 000 people are killed.
A–J below are responses to the disaster.

a Make a big table with headings like this:

Responses to the earthquake disaster	
Short-term	Long-term

b Now write out the sentences **A–J** in the correct columns. (With their label letters!)

A 30 firemen fly in from Italy to help.
B The government sets up a team of geologists to try to predict future earthquakes.
C A law is passed that all new buildings must be able to withstand shaking.
D All bridges in the city are made stronger.
E South American countries set up a joint satellite system, to monitor plate movements.
F From now on, all schools in the country will teach pupils what to do in an earthquake.
G A city hospital sets up a tent for the injured.
H Reporters arrive from all over the world.
I Spain gives $500 million to help rebuild the city.
J In Reading, year 7 students collect £80 for the survivors.

3 Which of the responses in your table for **2** are:
a local (in the city that was hit)?
b national?
c international?
To answer, underline each type of response in a different colour, in your table.
Then add a colour key.

4 A country's physical geography can make it hard to cope with a disaster.
This satellite image shows the area around Balakot in Pakistan, about 12 weeks after the earthquake. Only towns and larger villages hit by the earthquake are marked. Dozens of small villages are not shown.

When this image was taken, over 300 000 survivors were living in tents, or rough shelters. It was bitterly cold. Heavy snow had been falling for weeks. Thousands of people still had no help.

Using clues from the photo, suggest some reasons why it was hard to get help to people.

5 But there are other reasons a country can find it hard to cope with disasters. Look at the table below. It compares three countries that suffer earthquakes.
a Which one is wealthiest?
b Which one do you think is best able to:
i teach the public what to do in an earthquake?
ii get help to victims?
iii care for injured people?
Give your reasons for each answer.

6 *'It's harder for poor countries to cope well with disasters.'* Do you think this is true?
Give your reasons in not more than 40 words.

Comparing three countries

Country	Mexico	Pakistan	USA
Average wealth per person (in $US)	$9800	$2200	$40 000
TV sets per 1000 people	282	190	938
Number of doctors per 10 000 people	20	7	26
Number of hospital beds per 10 000 people	10	8	30
Length of motorway per 1000 sq km of the country	3.34 km	0.91 km	8.17 km

Ordnance Survey symbols

ROADS AND PATHS

M I or A 6(M) Motorway
A 35 Dual carriageway
A 31(T) or A 35 Trunk or main road
B 3074 Secondary road
Narrow road with passing places
Road under construction
Road generally more than 4 m wide
Road generally less than 4 m wide
Other road, drive or track, fenced and unfenced
Gradient: steeper than 1 in 5; 1 in 7 to 1 in 5
Ferry; Ferry P – passenger only
Path

PUBLIC RIGHTS OF WAY

(Not applicable to Scotland)

1:25 000 1:50 000

Footpath
Road used as a public footpath
Bridleway
Byway open to all traffic

RAILWAYS

Multiple track
Single track
Narrow gauge/Light rapid transit system
Road over; road under; level crossing
Cutting; tunnel; embankment
Station, open to passengers; siding

BOUNDARIES

National
District
County, Unitary Authority, Metropolitan District or London Borough
National Park

HEIGHTS/ROCK FEATURES

Contour lines

·144 Spot height to the nearest metre above sea level

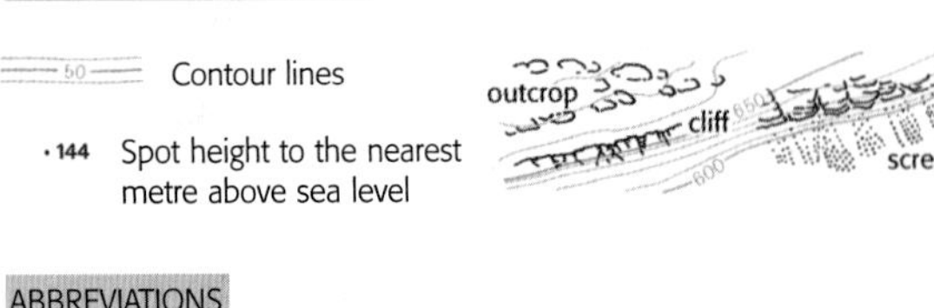

ABBREVIATIONS

PO	Post office	PC	Public convenience (rural areas)
PH	Public house	TH	Town Hall, Guildhall or equivalent
MS	Milestone	Sch	School
MP	Milepost	Coll	College
CH	Clubhouse	Mus	Museum
CG	Coastguard	Cemy	Cemetery
Fm	Farm	Hosp	Hospital

ANTIQUITIES

VILLA Roman
Castle Non-Roman
Battlefield (with date)
Tumulus

LAND FEATURES

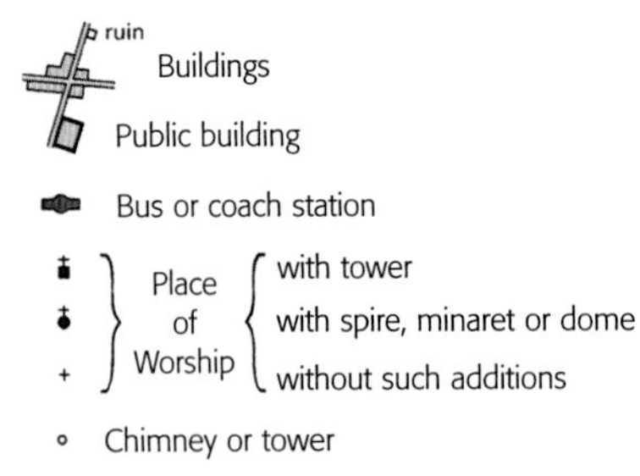

Bus or coach station

Place of Worship: with tower; with spire, minaret or dome; without such additions

Chimney or tower
Glass structure
Heliport
Triangulation pillar
Mast
Wind pump / wind generator
Windmill
Graticule intersection

Cutting, embankment
Quarry
Spoil heap, refuse tip or dump
Coniferous wood
Non-coniferous wood
Mixed wood
Orchard
Park or ornamental ground

Forestry Commission access land
National Trust – always open
National Trust, limited access, observe local signs
National Trust for Scotland

WATER FEATURES

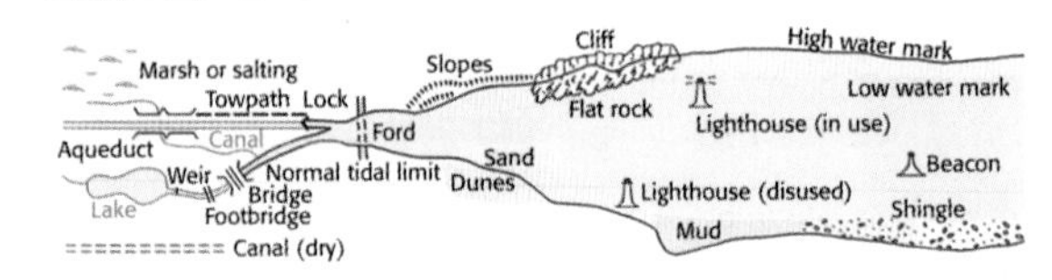

TOURIST INFORMATION

Parking
Visitor centre
Information centre
Recreation/leisure/sports centre
Telephone
Camp site/Caravan site
Golf course or links
Viewpoint
PC Public convenience
Picnic site
Pub/s
Cathedral/Abbey
Museum
Castle/fort
Building of historic interest
English Heritage
Garden
Nature reserve
Water activities
Fishing
Other tourist feature

Map of the British Isles

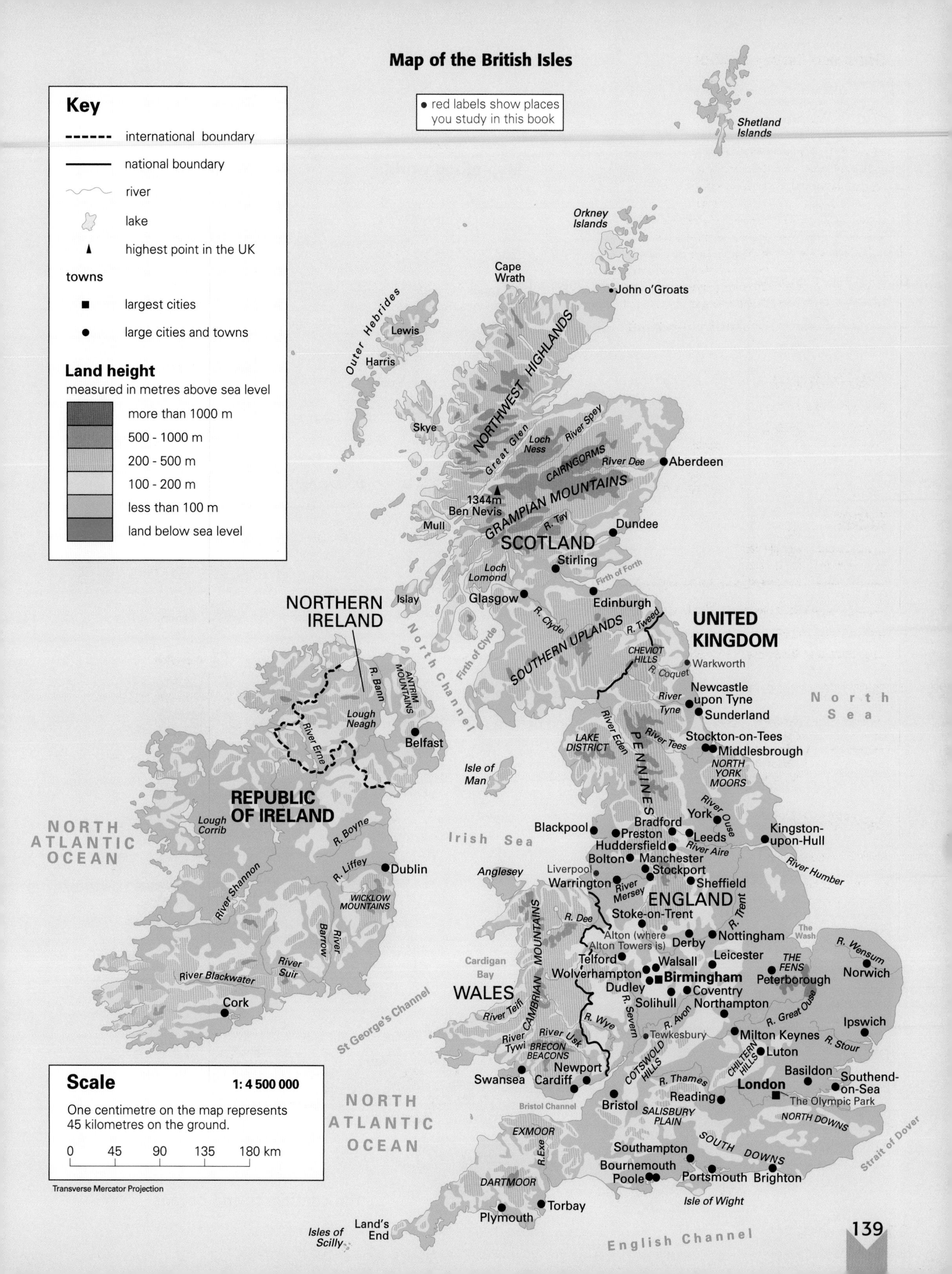

international boundary
capital city
abbreviations
BELG. BELGIUM
B-H. BOSNIA-HERZEGOVINA
C. CROATIA
CENT. AF. REP. CENTRAL AFRICAN REPUBLIC
CZ. CZECH REPUBLIC
F. FYROM
(Former Yugoslav Republic of Macedonia)
L. LIECHTENSTEIN
LITH. LITHUANIA
MT. MONTENEGRO
LUX. LUXEMBOURG
NETH. NETHERLANDS
S. SLOVENIA
SE. SERBIA
SL. SLOVAKIA
SWITZ. SWITZERLAND
U.A.E. UNITED ARAB EMIRATES
U.S.A. UNITED STATES OF AMERICA
Equatorial Scale 1: 95 000 000
Map of the world
Did you know ?
The Earth is 4600 million years old.
It weighs 6000 million million million tonnes.
Arctic Circle
Alaska (U.S.A.)
GREENLAND (Den.)
Nuuk (Godthåb)
CANADA
Ottawa
U. S. A.
Washington D.C.
Azores (Port.)
Bermuda (U.K.)
MEXICO
México
Tropic of Cancer
Hawaiian Is. (U.S.A.)
Nassau
THE BAHAMAS
Havana
CUBA
JAMAICA
Kingston
HAITI
DOMINICAN REPUBLIC
Puerto Rico (U.S.A.)
ST. KITTS-NEVIS
ANTIGUA & BARBUDA
DOMINICA
ST. LUCIA
BARBADOS
ST. VINCENT & THE GRENADINES
GRENADA
TRINIDAD AND TOBAGO
CAPE VERDE IS.
Belmopan
BELIZE
GUATEMALA
Guatemala
HONDURAS
Tegucigalpa
San Salvador
EL SALVADOR
NICARAGUA
Managua
San José
COSTA RICA
Panamá
PANAMA
Caracas
VENEZUELA
GUYANA
Georgetown
Paramaribo
SURINAME
Cayenne
FRENCH GUIANA (Fr.)
Bogotá
COLOMBIA
Equator
Galapagos Is. (Ec.)
Quito
ECUADOR
BRAZIL
PERU
Lima
Brasília
La Paz
BOLIVIA
Tokelau Is. (N.Z.)
American Samoa (U.S.A.)
SAMOA
French Polynesia (Fr.)
Niue (N.Z.)
Cook Is. (N.Z.)
TONGA
Tropic of Capricorn
Pitcairn Is. (U.K.)
PARAGUAY
Asunción
CHILE
ARGENTINA
Santiago
URUGUAY
Buenos Aires
Montevideo
Falkland Is. (U.K.)
Stanley
South Georgia (U.K.)
North America
Europe
Asia
NORTH ATLANTIC OCEAN
PACIFIC OCEAN
Africa
PACIFIC OCEAN
South America
INDIAN OCEAN
SOUTH ATLANTIC OCEAN
Oceania
SOUTHERN OCEAN
Antarctica
The continents and oceans

Amazing – but true!

- Nearly 70% of the Earth is covered by saltwater.
- Nearly 1/3 is covered by the Pacific Ocean.
- 10% of the land is covered by glaciers.
- 20% of the land is covered by deserts.

World champions

- Largest continent – Asia
- Longest river – The Amazon, South America
- Highest mountain on land – Everest, Nepal
- Highest mountain in the ocean – Mauna Kea, Hawaii
- Largest desert – Sahara, North Africa
- Largest ocean – Pacific

Did you know?

The world has:

- over 200 countries
- over 6 billion people
- over 6000 different languages.

Glossary

A

aftershocks – little earthquakes that follow the main one

aid agency – an organisation such as Oxfam that helps people in poorer countries

asylum seeker – a person who flees to another country for safety, and asks to be allowed to stay there

atmosphere – the layer of gas around the Earth

B

bedload – stones and other material that rolls or bounces along a river bed

brownfield site – a site that was built on before, but can be redeveloped

business park – an area planned and set up as a place for modern businesses

C

CBD – central business district – the area at the centre of a town or city where the main shops and offices are

Commonwealth – a 'club' of 53 countries, including the UK, that have historical links, and some shared goals; 51 of them were British colonies

commute – to travel to another place to your work (for example live in Aylesbury and commute to London)

comparison goods – things like clothes, shoes, and furniture, where you like to see a choice before you buy

condense – to change from gas to liquid

confluence – where two rivers join

contaminated – polluted by harmful things

contour line – line on a map joining places that are the same height above sea level

convection current – a current of warmer material; when air or water or soft rock is heated from below, the warmer material rises in convection currents

convenience goods – low-cost goods like milk, newspapers, and sweets, that you'll buy in the nearest convenient place

core – the inner layer of the Earth, made mainly of iron plus a little nickel

crater – a hollow around the vent of a volcano

crust – the thin outer layer of the Earth, made of rock

D

dam – a wall built across a river to control water flow

defence forces – the army, navy, air force; they defend the country from attack

deposit – to drop material; rivers deposit sediment as they approach the sea

derelict – run-down and abandoned

developers – companies that buy land and put up buildings, to rent or sell

development – a housing estate or shopping plaza or other group of buildings, built as one project

dormitory town – people live there, but travel somewhere else to their work

drainage basin – the land around a river, from which water drains into the river

dwelling – a building to live in (like a house or flat)

E

earthquake – the shaking of the Earth's crust, caused by sudden rock movement

economic – about money and business

economic activity – work you earn money from

economic migrants – people who move to a new place to find work, and to improve their standard of living

embankment – a bank of earth or concrete built up on a river bank, to stop the river flooding

emergency services – services such as the police, ambulance, and fire brigade, that help when people are in danger

emigrant – a person who leaves his or her own country to settle in another country

environmental – to do with our surroundings and how we look after them (air, rivers, wildlife, and so on)

epicentre – the point on the ground directly above the focus of an earthquake

equator – an imaginary line around the middle of the Earth (0° latitude)

erosion – the wearing away of rock, stones and soil by rivers, waves, wind or glaciers

European Union – a 'club' of European countries that trade freely with each other, and share many laws and aims; 27 countries are in it (including the UK)

evacuate – move from a dangerous place to a safe one

evaporation – the change from liquid to gas

F

fault – a crack in the Earth's crust, where rock has moved

flash flood – a sudden flood usually caused by a very heavy burst of rain

flood – an overflow of water from the river

flood defences – structures built to prevent flooding; for example an embankment (raised river bank) along the river

flood plain – flat land around a river that gets flooded when the river overflows

focus – the 'centre' of an earthquake; it is the place where the rock moved

fold mountains – mountains formed as a result of plates pushing into each other

fossil fuels – coal, oil, and natural gas

function – an activity in a settlement; its main funtion could be as a port, or holiday resort, for example; but other functions go on too (such as shopping)

G

geologist – a scientist who studies rocks, earthquakes, and so on

glacier – a river of ice

global warming – average temperatures around the world are rising; experts say this is mainly due to carbon dioxide from burning fossil fuels

gorge – a narrow valley with steep sides

greenfield site – a site that has not been built on before

groundwater – rainwater that has soaked down through the ground and filled up the cracks in the rock below

H

high-value business – provides goods or services that people will pay a lot for

I

immigrant – a person who moves here from another country, to live

impermeable – does not let water pass through

Industrial Revolution – the period of history (around the 18th century) when many new machines were invented and many factories built

industry – a branch of manufacturing or trade, such as the car industry

infiltration – the soaking of rainwater into the ground

infrastructure – the basic services in a country, such as roads, railways, water supply, telephone system

insurance – you pay a fee to insure an item; then if it gets lost or damaged, the insurance company gives you money

interdependence – how people, or countries, depend on each other

international – to do with more than one country

internet – a network of millions of computers around the world, all linked together

invader – enters a country to attack it

irrigate – to water crops

L

landfill site – a place set aside as a big rubbish dump

landform – a feature formed by erosion or deposition (for example a gorge)

lava – melted rock that erupts from a volcano

lithosphere – the hard outer part of the Earth's surface; it is broken into big slabs called plates

local – to do with the area around you

long-term – for the years ahead, stretching into the future

M

magma – melted rock below the Earth's surface; when it reaches the surface it is called lava

magnitude – how much energy an earthquake gives out

mantle – the middle layer of the Earth, between the crust and the core

manufacturing – making things in factories (in the secondary sector of the economy)

market town – a town that grew because of its market (like Aylesbury did); today it may have some industry too

meander – a bend in a river

media – forms of communication, such as TV, radio, newspapers, the internet

mental map – a map you carry in your head

merchandise – goods for sale

migrant – a person who moves to another part of the country, or another country, often just to work for a while

mudflow – a river of mud; it can form when the material from an eruption mixes with rain or melting ice

N

national – to do with the whole country (for example the national anthem)

NGO (non-governmental organisation) – a not-for-profit organisation, such as Oxfam, that helps people, and is independent of the government

North Atlantic Drift – a warm current in the Atlantic Ocean; it keeps the weather on the west coast of Britain mild in winter

O

oxbow lake – a lake formed when a loop in a river gets cut off

OS maps – very detailed maps, drawn to scale, by the Ordnance Survey

P

permeable – lets water soak through

persecute – to punish or treat cruelly (for example because of race or religion)

plan – a map of a small area (such as the school, or a room) drawn to scale

plates – the Earth's surface is broken into large slabs; these are called plates

plunge pool – deep pool below a waterfall

population density – the average number of people per square kilometre

precipitation – water falling from the sky (as rain, sleet, hail, snow)

prevailing winds – the ones that blow most often; in the UK they are south west winds (they blow *from* the south west)

primary sector – the part of the economy where people take things from the Earth and sea (farming, fishing, mining)

pyroclastic flow – a flood of gas, dust, ash and other particles rushing down the side of a volcano, after an eruption

Q

quaternary sector – the part of the economy where people do high-tech research (for example into genes)

R

redevelop – to rebuild an area for a new use

refugee – a person who has been forced to flee from danger (for example from war)

regenerate – redevelop a run-down area and bring it back to life again

residential area – an area which is mainly homes (rather than shops or offices)

Richter scale – a scale for measuring the energy given out in an earthquake

Ring of Fire – the ring of volcanoes around the Pacific Ocean

rural area – countryside, where people live on farms and in small villages

S

scale – the ratio of the distance on a map to the real distance

secondary sector – the part of the economy where people make or build or process things (such as cars, furniture, food)

sediment – a layer of material (stones, sand and mud) deposited by a river

seismic wave – wave of energy given out in an earthquake; it shakes everything

seismometer – an instrument for recording the vibrations during an earthquake

settlement – a place where people live; it could be a hamlet, village, town or city

settler – a person who takes over land to live on, where no one has lived before

sewage works – where all the waste liquid from our homes is cleaned up, before it is put back in the river

short-term – for the days and weeks ahead

site – the land a settlement is built on

sketch map – a simple map to show what a place is like, or how to get there; it is not drawn to scale

social – about society and our way of life

source – the starting point of a river

sphere of influence – the area around a settlement (or shop, or other service) where its influence is felt

spot height – the exact height, in metres, at a spot on an OS map (look for a number)

suspension – small particles of rock and soil carried along in a river; they make the water look cloudy or muddy

sustainable – brings economic, social and environmental benefits; can be carried on into the future, without doing harm

T

tertiary sector – the part of the economy where people provide services (for example teachers, doctors, taxi drivers)

TNC (transnational corporation) – a big company that has branches in different countries; McDonald's is an example

trench – a deep steep valley in the ocean floor

tributary – a river that flows into a larger one

tsunami – a series of giant waves set off by an earthquake in the ocean floor

U

urban area – a built-up area, such as a town or city

urban regeneration – when a run-down urban area is redeveloped and brought to life again

U-shaped valley – a valley shaped like the letter U, carved out by a glacier

V

vale – a very broad open valley

valley – an area of low land, with higher land on each side; it was carved out by a river or glacier

vent – a hole through which lava erupts, on a volcano

venue – a place where an event is held; for example, Wembley Stadium

volcano – a place where lava erupts at the Earth's surface

vulcanologist – a scientist who studies volcanoes

V-shaped valley – a valley shaped like the letter V, carved out by a river

W

water cycle – the non-stop cycle in which water evaporates from the sea, falls as rain, and returns to the sea in rivers

water vapour – water in gas form

waterfall – where a river or stream flows over a steep drop

watershed – an imaginary line separating one drainage basin from the next

weather – the state of the atmosphere – for example how warm or wet it is

weathering – the breaking down of rock, caused mainly by the weather; it turns into soil in the end

Index